Luxury, Fashion and the Early Modern Idea of Credit

Luxury, Fashion and the Early Modern Idea of Credit addresses how social and cultural ideas about credit and trust, in the context of fashion and trade, were affected by the growth and development of the bankruptcy institution.

Luxury, fashion and social standing are intimately connected to consumption on credit. Drawing on data from the fashion trade, this fascinating edited volume shows how the concepts of credit, trust and bankruptcy changed towards the end of the early modern period (1500–1800) and in the beginning of the modern period. Focusing on Sweden, with comparative material from France and other European countries, this volume draws together emerging and established scholars from across the fields of economic history and fashion.

This book is an essential read for scholars in economic history, financial history, social history and European history.

Klas Nyberg is Professor of Fashion Studies at Stockholm University, Sweden.

Håkan Jakobsson is a PhD student in the Department of History at Stockholm University, Sweden.

Perspectives in Economic and Social History
Series Editors: *Andrew August and Jari Eloranta*

For more information about this series, please visit www.routledge.com/series/PESH

Luxury, Fashion and the Early Modern Idea of Credit

**Edited by
Klas Nyberg**

**Co-edited by
Håkan Jakobsson**

Routledge
Taylor & Francis Group

LONDON AND NEW YORK

First published 2021
by Routledge
2 Park Square, Milton Park, Abingdon, Oxon OX14 4RN

and by Routledge
52 Vanderbilt Avenue, New York, NY 10017

Routledge is an imprint of the Taylor & Francis Group, an informa business

British Library Cataloguing-in-Publication Data
A catalogue record for this book is available from the British Library

Library of Congress Cataloging-in-Publication Data
Names: Nyberg, Klas, editor. | Jakobsson, Håkan, editor.
Title: Luxury, fashion and the early modern idea of credit / edited by Klas Nyberg and Håkan Jakobsson.
Description: Abingdon, Oxon ; New York, NY : Routledge, 2021. | Series: Perspectives in economic and social history | Includes bibliographical references and index.
Identifiers: LCCN 2020029446 (print) | LCCN 2020029447 (ebook) | ISBN 9780367332693 (hardback) | ISBN 9780429318979 (ebook)
Subjects: LCSH: Consumption (Economics)—Social aspects—Europe—History. | Luxuries—Europe—History. | Credit—Social spects—Europe—History. | Fashion—Europe—History.
Classification: LCC HC240.9.C6 L89 2021 (print) | LCC HC240.9.C6 (ebook) | DDC 306.3094—dc23
LC record available at https://lccn.loc.gov/2020029446
LC ebook record available at https://lccn.loc.gov/2020029447

ISBN: 978-0-367-33269-3 (hbk)
ISBN: 978-0-429-31897-9 (ebk)

Typeset in Times New Roman
by codeMantra

Contents

Figures

Tables

Contributors

Marcus Box Associate Professor in Economic History and currently employed as senior lecturer in Business Studies at Södertörn University. His research interests mainly concern entrepreneurship and firm dynamics, such as self-employment, bankruptcies and the entry, growth and exit of firms.

Karl Gratzer Professor Emeritus in Economic History at Södertörn University, where he was leader of research and higher education at the Institute for Business Administration. His research areas include the interface between economic history, business administration and economics with a focus on public and private entrepreneurship, small business development, insolvency issues and questions on selection processes within clusters of firms. Gratzer is one of the founders and a board member of the Swedish Centre for Research on Insolvency and a member of the editorial board for the *Journal for Insolvency Law*.

Axel Hagberg PhD and Researcher in Economic History at the Stockholm School of Economics, affiliated with the Institute for Economic and Business History Research and the Department of Economic History at Uppsala University. His research focus is on financial history.

Mats Hayen PhD in History from Stockholm University and Research Manager at the Stockholm City Archives. His research areas include urban history, urban geography, time geography, economic history, social history, children's history, archival databases and population history.

Håkan Jakobsson PhD student in History at the Department of History and also active at the Department of Media studies/Centre for Fashion Studies at Stockholm University. His research areas include commerce, technology, networks, bankruptcies, consumption and the role of knowledge and expertise in the early modern period.

Xiang Lin PhD and Senior Lecturer in Economics at Södertörn University. His research interests are in the field of macroeconomics and empirical analyses, with a particular focus on entrepreneurship.

Klas Nyberg Professor of Fashion Studies at the Department of Media Studies/ Centre for Fashion Studies at Stockholm University, where he has served as Director and now heads the postgraduate research programme. He was previously Professor of Economic History at the Department of Economic History at Uppsala University. His dissertation was awarded the Uppsala University Geijer Prize in 1995. In 2013 he was awarded the Söderberg Prize and in 2018 the St Erik Medal from the City of Stockholm. His research areas include early modern social and economic history, textile and financial history.

Riina Turunen PhD and Postdoctoral Researcher at the University of Jyväskylä. Her research focus is on pre-industrial economic growth, industrialization, currency conditions, as well as early modern credit markets and the role of women in insolvency cases.

Göran Ulväng Researcher and Associate Professor at the Department of Economic History at Uppsala University. His research focus includes changes in agriculture, buildings, households, credits and insurance practices in eighteenth and nineteenth century Sweden, with a special emphasis on manors and manorial culture.

Paula von Wachenfeldt Associate Professor in Fashion Studies at the Department of Media Studies/Centre for Fashion Studies at Stockholm University. Her research addresses, among other things, fashion representations in literature and fashion communication. She has also devoted her research to luxury studies where she has addressed the debate on luxury and the interpretation of luxury in media and advertisement. Her research field covers both historical and contemporary controversies.

Kustaa H. J. Vilkuna Professor of Finnish history at the University of Jyväskylä. His research is focused on early modern cultural history, bourgeois consumption, political fashion, wigmakers in the Nordic countries and drinking culture.

Acknowledgements

The research presented here has been funded by the Torsten Söderberg Foundation. The work has been conducted at the Department of Media Studies/Fashion Studies, Stockholm University; the Institute for Economic and Business History Research (EHFF), the Stockholm School of Economics; and the Stockholm City Archives. I extend my thanks to the institutions concerned for their courtesy and support of the project. I would also like to thank the participating researchers and, in particular, my co-editor Håkan Jakobsson for his dedicated work. Thanks also to Professor Emeritus Håkan Lindgren and Associate Professor Anders Perlinge at EHFF and Associate Professor Elise Dermineur at Umeå University for the opportunity to present the project at the conference *Wealth and Debt Accumulation in Early Financial Markets. A Marcus Wallenberg Symposium* at the Stockholm School of Economics in September 2018. Thank you as well to Professor Heiko Droste, Department of History, Stockholm University who chaired a session about the project, "Fashion, Luxury, Credit and Trust. Early-Modern and Modern Bankruptcies in Europe, North America or Elsewhere" at the *European Social Science History Conference*, at Queen's University Belfast in April 2018. Finally, a thank you to Professor Johan Fourie, Department of Economics, Stellenbosch University, for the opportunity to finish the anthology in a welcoming and stimulating academic environment during the spring of 2020 despite the strains of the COVID-19 pandemic.

Stockholm, September 2020,
Klas Nyberg

General introduction

Klas Nyberg

Introduction

During the eighteenth century a vast private credit market built up around
unsecured short-term loans, where credit was connected to personal trust
rather than to collateral such as property or other assets, started to emerge
in Europe. The development was closely connected to a growing demand for
consumption of goods and luxurious consumption that happened across the
continent during this period. In order to respond to this growing demand,
more working capital was needed than before among traders and producers,
who were dependent on private lenders to secure credits. Commercial
banks and other institutional lenders only became important after the
mid-nineteenth century. This anthology addresses some of the problems
that grew out of the practice, and in particular how it affected urban society
when burghers, business owners and other members of society increasingly
expanded beyond their means and ended up in financial difficulties.[1]

An increasing demand for various new consumption goods can be
established in the Western World from the seventeenth century, with a starting point in Northwestern Europe. The consumer-driven development has
been characterized by the economic historian Jan de Vries as an industrious
revolution, a process that predated the industrial revolution. His theoretical
approach is based on the observation that everyone in the early modern
household, including women and children, started to work more than before in order to purchase a wider and growing number of new luxury and
fashionable items, primarily colonial products and textiles. The new approach suggests the need to view the early industrialization as a process
that contained a strong dynamic element of consumerism.[2]

Less recognized in this context is how this type of growing consumption
of new goods was connected to or potentially even the result of the rise of
a new private credit market centred on the use of promissory note. These
types of loans without securities ranged from small to very substantial.
They were used for a wide range of things related to the personal sphere,
including investments in permanent displays of opulence such as buildings
or the purchase of durable fashionable consumer goods, but also for the

consumption of perishable goods. In this anthology, the increasing ability to secure promissory notes is viewed as a fundamental factor behind the growing consumption of modern and fashionable items and generally to a new lifestyle based on conspicuous consumption.

The anthology is set against a theoretical framework which is founded in the various ways in which the concept of credit changed towards the end of the eighteenth and in the beginning of the nineteenth century.[3] In the various contributions this can be concretely seen in a particular aspect of consumption and economic strategies, namely the managing of debts and survival strategies in situations of economic turmoil and personal insolvency. This, in turn, relates to the creation of a legislative framework to handle debt and economic crisis as part of a wider state activity coupled to the realization about the need for growing economic realism at the dawn of the modern era.[4]

This modernization process was not unique to Sweden but happened in many European countries. In Sweden, the new legislation crucially included the creation of a new bankruptcy legislation that was implemented between 1767 and 1818. The new legislation changed the way in which bankruptcies were managed, which led up to a more structured and time-efficient system, where the average time from application to verdict in a bankruptcy case was substantially shortened.[5]

The growing focus on efficiency, structure and general economic management on the part of the state can be contrasted to a continued lack of economic realism among many groups in society.[6] How this played out in reality, over time among various groups, in various industrial or handicraft sectors and across social boundaries in Swedish society is at the core of the various contributions in this anthology. By focusing on the economic strategies and considerations in the bankruptcies that happened during the period, set against a backdrop where the state attempted to reform the financial legislation and related to the theoretical ideas of a new economic behaviour, a new view of economic behaviour and strategies at the onset of the modern period will be provided.

Credit, trust and bankruptcies

Concepts such as credit, security, property and dowries had more than strictly economic functions during the early modern period.[7] Financial affairs that were undertaken by business owners and artisans were embedded in social networks, based on family, gender, ethnicity and confession.[8] For a merchant, a career as an independent burgher was often only begun after a long period of practical training as a trade apprentice or commercial clerk in one or more commercial offices. This was not just a learning period; it was also a period for the person in question to build up his creditworthiness. At the end of this period when applying for burgher rights his creditworthiness was guaranteed by special so-called "sponsors", commonly two burghers of good repute.[9]

Creditworthiness had a wider significance than sound business sense, solvency and good liquidity. According to economic historian Laurence Fontaine, in the eighteenth century, the French word *crédit* meant that a person acquired respect through, for example, public office, and had integrity and was honest. It was also linked to a person's reputation based on prosperity, power and authority.[10] Many researchers have shown the existence of and stressed the absolute necessity of extensive credit networks as a central part of early modern wholesale trade and as an important part of the basic development conditions of medieval and early modern entrepreneurship.[11]

Credit relations in early modern times relied on personal networks rather than on banks and other credit institutions. The complicated network relations between private actors, bankers, kinsmen, wholesale and retail traders, artisans and institutions all tended to be based on trust. To have a good credit rating meant that others trusted in you as a person, and your trustworthiness as a business owner, in turn, came from the various networks in which you participated.[12]

Large parts of the assets of wholesale traders, manufacturers and craftsmen took the form of claims on other individuals or companies. If the owner of a business operation lost his creditworthiness or filed for bankruptcy, other actors' claims on his estate were soon in doubt and were in risk of being regarded as worthless assets. As soon as a bankruptcy was announced, good debts became bad debts.[13]

This had serious implications beyond the impact on the creditors' economies. The faith and personal trust between the parties or essentially the social underpinnings that supported the credit were gone and in many cases replaced by hatred and desperation. In Stockholm this can be seen in thousands of preserved bankruptcy cases from the eighteenth century. In detailed submissions to the municipal court, handed in at the beginning of the bankruptcy proceeding, former business associates or members of shared social networks suddenly request priority or even that the debtor should be jailed. Behind the sometimes theatrical tone in the documents lay the very real fact that the creditors not only tried to safeguard their economies, but more important still, their social positions. The failure of the debtor in short meant that the credit of his business partners came under threat.

In Sweden, at the onset of the modern era, the state took on an important new role as the driver behind the modernization of the credit market. One of the results of this development was a formalization and rationalization of the bankruptcy institution. Before the eighteenth century, the procedures to handle insolvencies were limited in scope and efficiency in Sweden, as in many European countries.[14] The introduction of new laws and additions to existing laws, in the Swedish case, led to the shortening of the time between application and verdict and generally to a growing number of rulings.

This not only provided for faster administration but also crucially played an important role in the reconstruction of credit networks and companies.[15]

The legislation was designed to safeguard the assets in the bankrupt estate, divide them and ensure that none of the creditors were allowed to profit at the expense of the others through compulsion and manipulation. With the new legislation the state challenged early modern ideas about the right to credit based on personal principles, social status or prominent societal positions. This included such notions that it was acceptable to disregard book-keeping or sound financial management. Another notion that was challenged was the idea that a husband or guardian was entitled to use dowries or inheritances.

During the early modern period traders and manufacturers transferred capital to each other's businesses via daughters' dowries.[16] While a kind of underlying taboo, which prevented men from plundering their wives' dowries as well as their children's inheritance, was present, few legal ramifications for the behaviour existed. In Sweden this all changed with the introduction of the new legislation. By the end of the eighteenth century, married women in Sweden could refer to the relevant section of the commercial code, the bankruptcy law and the marriage code, which combined offered legal protection for their property in the event of a bankruptcy.[17] In practice this meant that women with dowries or other means to protect often sought a judicial separation of the estate with a request to be freed from responsibility for their spouses' debts once a bankruptcy was a reality. This was also a common claim from women whose spouses had abused their guardianships or made decisions about real estate without the consent of their wives.[18]

The industrious revolution and new luxury

Jan de Vries in 1994 coined the term, the industrious revolution, to describe a new type of consumer behaviour which he observed in Western European societies from the latter part of the seventeenth century onwards.[19]

His approach, as mentioned initially, theorizes how a growing demand for new consumer goods can be understood as a product of changing resource allocations in the early modern household. While partly basing his ideas on Akira Hayami's observations about early modern Japan, he also fell back on more empirically designed approaches from the 1980s on a consumption revolution, formulated in the anthology "The Birth of a Consumer Society" and "Consumption and the world of goods".[20]

The idea about an industrious revolution more precisely is based on the observation that all members of many early modern households during the long eighteenth century started to work more in order to purchase a wider and growing number of new luxury and fashionable items, primarily colonial products such as coffee, tea, tobacco, sugar as well as textiles. The new consumption happened despite a general rise in the price of food, an expense which constituted more than half of the total income for most households, and despite a decrease in real wages as a whole during the period 1650–1800. At the same time, the prices of certain goods, such as linen and sugar, saw a strong decline relative to basic foodstuff. The new approach suggests the need to view the early industrialisation as a process that contained a strong dynamic element of consumerism.

Researchers such as Shaohua Zhan, Craig Muldrew, Ragnhild Hutchison and Akira Hayami have in recent years further problematized the theory by using it to interpret the development in different parts of the world. As a result, new knowledge of development patterns related to the theory has been unearthed in Asia (Japan and China), in Northwestern Europe (Great Britain) and in Scandinavia (present day Norway).[21] The authors have provided new results that have toned down the idea that the demand for consumer goods was the key reason why households sought to participate in the market economy. They have at the same time provided information which can be used to show interesting differences between major countries in Asia and Europe, and for example the Scandinavian periphery.

Hayami's and Zhan's studies suggest that the Asian move towards an industrious revolution was based on changes in the organization of agriculture. Hayami suggests that a large part of the population of early modern Japan became industrious because of the emergence of a labour intensive agriculture that happened at the same time as the emergence of an increased tendency towards involvement in a market economy. Zhan's book on the rise of capitalist agriculture in China also links the labour intensive nature of the development to a more general agrarian transformation. Finally, on the basis of extensive source studies of the theory's geographical core area: Northwestern Europe, Muldrew argues that households in this region did not become more industrious solely because of changing preferences. The process here occurred at the same time as changes in the labour market and new policies that encouraged work.[22]

The changes in early modern Asia as well as the conditions in Great Britain before 1780 initiated an industrious revolution. It also characterized the social and economic transformation process in Scandinavia that began in the early 1800s, some 40–50 years after the British development. In the sparsely populated Norway (which was part of Denmark until 1814), the industrious revolution happened in interaction with an emerging export industrialization according to Hutchison. When timber and other staple goods received an increased economic value as export commodities, the household incomes and their understanding of import goods grew. Hutchison emphasizes that the household strategies shifted over business cycles and how they were forced to prioritize food during difficult times. This also suggests that there was no unilateral link between industriousness and changing preferences in the peripheries.[23]

Another theory which also builds on the same earlier research discussions about a consumer revolution has been proposed by Maxine Berg. Much like de Vries she has also drawn attention to shifting patterns of consumption and more specifically to the rise of a new type of luxury consumption in the eighteenth century.[24] The idea has been referred to as new luxury. The concept includes materials and commodities that were made available through global trade networks, but also the very process, through which such imports influenced and affected the production of imitations in domestic manufactures.[25] Crucially, it includes a wider range of commodities and goods than old luxury: the more traditional luxury consumed by the aristocracy and the higher ranks of society.

New luxury as a changing, modern concept based on the use of novel commodities was closely connected to the rise of fashion, which was shaped by social interactions in towns, where, according to Berg, the growing middle class used information, style and taste to embrace a life of modernity. New luxury was not merely created by producers but also by merchants, advertisers and retailers. It was in this sense as much the introduction of new merchandise as it was the feeling for the relationship between the new products, the knowledge of their existence and qualities and the ability to use and display them as part of one's person. This for example included things like how to arrange the interior in a new residence with the correct colours, textiles and styles of furniture.[26]

Overall the ideas about consumerism and luxury discussed here form part of a broader notion of social position and social rights as a product of economic behaviour.[27] Whereas new luxury primarily was consumed by the middle class, with merchants, industrialists and artisans at the forefront, old luxury was synonymic with traditional aristocratic consumption. It epitomized extravagant opulence, often inherited or handed down through the generations, which mainly existed because of the possibility to engage in wasteful spending on credit. This spending was based on a politically favoured position including substantial tax-exemptions. Next to luxurious and fashionable objects it was also displayed through a grand lifestyle in castles and mansions. Old luxury was prominently addressed in the sumptuary laws of the early modern period, which tried to preserve status quo and avert a broader consumption of such traditional luxury.[28]

A growing credit market for promissory loans

There is a benefit in combining both theories. The new consumer behaviour that first emerged in the late seventeenth century and the new type of historical dynamism that followed in its wake can be seen as the necessary underpinnings for the ideas about a new luxury consumption that took hold in the eighteenth century. What is lacking however are the financial aspects that can explain how this new consumer behaviour was made possible.

Maxine Berg acknowledges the problem when summarizing her perspective towards end of her book *Luxury and Pleasure*: "Luxury and Pleasure has attempted to restore something of that unity – goods, work, and consumption – in its framework of conception, making, and shopping.

But it investigates only a very few consumer goods in any depth, and neglects the social framework of credit and debt that made the making and buying of these goods possible".[29]

The proposal here is that the demand for new luxury and the related consumption of fashion goods increased the need for short-term unsecured loans or in other words short credits. This challenged the existing credit market, which was built up around long-term credit arrangements, based on collateral in property.

The latter type of arrangement was typical for aristocratic owners of manors, ironworks and other production facilities, who sold their production through merchants. In return the merchants provided both luxury and fashion goods as well as essential items on credit based on the projected future returns from the production. Once a year, the ratio of profit to return was regulated in relation to the cost of the consumer goods. If the nobleman consumed beyond his means, the deficit was recorded in the books and formed the starting point for the next financial year. If the deficit grew and became permanent to maintain a conspicous lifestyle, the owner risked losing all or part of his property to the merchant.[30]

While old luxury was basically building slow predictable consumption, new luxury was something new. The consumption of a wider range of luxury goods was, as mentioned, fundamentally made possible by an increased industriousness. From a financial perspective, it was enabled by short credits, small loans taken without collateral.[31]

In this new context, things like fabrics, clothing, fashion accessories, furniture and home furnishings, as well as various kinds of foodstuff and perishable products, were paid with promissory notes. Whereas the creditworthiness of a nobleman was guaranteed by collateral in the form of real estate, borrowing on the market for promissory notes was based on an individual's creditworthiness. A person was creditworthy, not only based on his financial strength but also based on his social position and the cultural conceptions of the conduct of a person of his standing.

Indebtedness was often only discovered in connection when the estate was divided after the death of a person, or in more obvious cases, when the person applied for new loans or tried to convert old ones. If such situations did not occur, a person could have hundreds of promissory notes in circulation on the credit market and technically be insolvent for several years, but still maintain his lifestyle and social function.

The rise of increased consumption of fashion goods and new luxury was thus intimately associated with the growing private reverse loan market. A key underlying aspect in the anthology is that the reformation of the bankruptcy institution was an important element in tightening the rules governing the private credit market for short-term reverse loans with an increased economic realism as a likely consequence. This assumption is supported by the extensive research that followed de Vries's first article on the concept of the industrious revolution, namely that borrowing grew in scale and significance in the light of the growing demand for fashion and luxury goods.[32] The growing incomes that followed the industrious behaviour were, in short, not enough.

With a new credit behaviour, followed a growing number of bankruptcies, and thus a need to reform the institutions to handle the new situation and encourage the development of economic realism.

In Sweden this not only affected the bankruptcy institution but also led to attempts to reform and to modernize the monetary system and currency market 1834.[33] The credit market of the late eighteenth century still contained elements of a subsistence economy where goods that were deliveries were stated in monetary equivalents but were paid in kind. Furthermore, the financial system had different types of parallel currencies that were not fully exchangeable, where loans often had to be repaid in the original currency. To finance wars, kings and princes often undermined fiscal reforms by creating temporary paper currencies without financial coverage.[34]

Summary

In sum, the objective of this anthology is to discuss and problematize the transformation of early modern ideas of credit and fashion among luxury producers like manufacturers and artisans and merchants and how credit and trust among those groups were affected by a Stockholm's demographic and economic stagnation and the reformation of the bankruptcy institution from the end of the eighteenth century. This is set against a theoretical framework that suggests that the luxury and fashion and the assertion of rights for different groups in society were intimately connected to consumption on credit.[35]

The contributors focus specifically on how businesses and activities in the fashion and luxury industries in Sweden related and functioned in relation to the state-supported modernization of the bankruptcy institution. Did they exhibit a more modern economically rational behaviour over time or was an early modern broader view of credit maintained despite the new state policies? Did traditional ideas about the right to consume and invest without financial coverage persist, did a new modern behaviour influence decision making or were perhaps alternative strategies that attempted to reconcile the two approaches attempted?

The structure of the book

In Part I focus is on the French influence on early modern fashion, both in a Parisian context, and how it was used in Stockholm and Sweden in the first half of the eighteenth century. The approach builds on the commonly held view that fashion in Europe was a development with roots in a French context, essentially as something that spread from Versailles and Paris to other parts of Europe in the early modern period.[36] In the first chapter, the discursive level is dealt with a focus on how fashion was presented in seventeenth- and eighteenth-century Paris, in literature and dramas as well as in fashion magazines. In the following chapter, focus is on the impact which French fashion had in Sweden during the first half of the nineteenth century in fashion magazines and the daily press.

In Part II focus is on the bankruptcy institution as a crucial feature in the modernization of the Swedish economy. The first chapter deals with the development of the Swedish bankruptcy institution from the eighteenth and first half of the nineteenth century against a historical as well as a European background. In the second chapter a novel approach to determine the bankruptcy frequencies and the structural development of insolvency in Stockholm and Gothenburg is presented.

In Part III focus is on the luxury and fashion industry in Stockholm, with different chapters covering specific fields, with studies of silk weaving, book printing, wigmaking, furniture making and painting. The first chapter provides a broader outlook of Stockholm's institutional, demographic, economic and social structure, viewed in an international comparative perspective. It also introduces the idea about contact zones as a way to understand how new knowledge was created in urban centres by bringing together historically and geographically separated groups in new ways.[37]

In Part IV the studies are summarized and the results are compiled and discussed in an international perspective in relation to the theoretical concepts.

Part I: Paris: the capital of luxury

Rational follies: fashion, luxury and credit in eighteenth-century Paris

This chapter examines the history of ideas behind the establishment of fashion and luxury as a social practice in pre-Revolutionary France and its reliance on the credit system. By examining different writings like popular plays, memoirs and fashion magazines from the time, the chapter shows that credit connected to sartorial consumption was the fruit of social ideas and cultural beliefs about the deeply entrenched role of clothing in society. The chapter highlights the role played by *marchandes de modes* as individuals who gave rise to both the culture and economy of fashion. It furthermore underscores their contribution to the transformation of old luxury as a fundamental feature of consumption for the elite classes to new luxury as a fashionable and changeable state of consumption for both the elite and middle classes. Finally, it shows how the establishment of the idea of fashion as a commercial business in eighteenth-century France relied on credit systems. This laid the grounds for the perpetuation of fashion and luxury as a sartorial and cultural business in future societies.

The French model and the rise of Swedish fashion, 1800–1840

This chapter investigates the impulses for extravagant consumption of fashion and luxury on credit in Sweden in the early nineteenth century through an investigation of the role of French-inspired fashion magazines and the advertisement in newspapers. The first fashion magazines in Sweden were largely translations of foreign magazines that were completely dominated by fashion and news from Paris. Almost three-quarters of these magazines were published in Stockholm. The chapter discusses how the publication and the contents of these magazines together with French-inspired fashion in advertising in newspapers affected Swedish fashion and consumption.

Part II: The Swedish financial system and bankruptcy law

The Swedish bankruptcy system, 1734–1849

This chapter details how new procedures were established in the handling of bankruptcies in Stockholm between 1734 and 1849. To minimize the damage from personal bankruptcy, an institutional administration of the bankruptcy developed across much of early modern Europe. The development is viewed as part of a historical and European context and discussed against the backdrop of the demographic and economic stagnation of the Swedish capital. The new legislation changed the way in which bankruptcies were managed, leading up to a more structured and time-efficient system, where the average time-frame from application to verdict in a bankruptcy case was substantially shortened.

Bankruptcies in Sweden, 1774–1849. Causes and structural differences

This chapter uses demographic, financial and economic variables, along with broader structural descriptions to explain variations in the volume of bankruptcies in Gothenburg and Stockholm, during the period between 1774 and 1849. Historical research on bankruptcies has been dominated by case study approaches that are often not suitable for generalization. The results from case studies are sometimes anecdotal and most often not statistically generalizable. Furthermore, micro-oriented explanations of the causes for bankruptcies, such as incompetency or fraudulent behaviour, can also be reductionist. Often, the dominant explanation for bankruptcies during the eighteenth and nineteenth centuries stresses various personal vices or corrupting tendencies or extensive speculation. This viewpoint can also be seen reflected in the contemporary legislation. These largely individual-based and moralizing explanations can be questioned. The present study instead approaches the problem through statistical analysis. The study is based on more than 20,000 bankruptcies as well as collections of demographic and financial statistics to compare bankruptcies in the two towns.

Part III: Credit and bankruptcies in the fashion and luxury trades in Sweden, 1730–1850

The institutional setting of the luxury trades in eighteenth and early nineteenth-century Stockholm

This chapter discusses the formation of the fashion industry and luxury artisanal crafts in Stockholm from the 1720s until the 1760s and thus serves as a background to the case studies that follow in the rest of Part III. The development is carried forward and presented in the light of the long-term impact of the demographic and economic stagnation that characterized the city after from the 1760s until the mid-nineteenth century. This latter period was characterized by weak population growth, high mortality rate, falling marriage rates and an increase in the number of extra-marital children. In economic terms, Stockholm as a trading town and the textile manufacturing industry was de-industrialized after 1760. Sharply declining real wages, declining living standards with increasing poverty were combined with increasingly unequal income distribution. Stockholm was among a group of cities that included a number of cities in and around the Mediterranean that showed the weakest population growth in Europe.

Economic behaviour and social strategies in the Stockholm silk weaving industry, 1744–1831

This chapter investigates the Stockholm silk weaving industry from the 1740s until the early 1830s. The historical background and the rise of the industry and its French influence are first discussed. This is followed by an analysis of the long-term development and subsequent decline of the industry based on industrial statistics, as part of a discussion on ways in which the industry coped during periods of economic crisis. Finally the chapter provides a new insight into the way in which individual silk weaving firms survived over time, and how bankruptcies were used or avoided in the sector. This is done through an in-depth discussion of the social networks that permeated the industry and by analysing the role of one of the major capital investments in the silk-weaving enterprises, namely the looms on which the actual weaving took place.

Hair professionals in financial distress in Stockholm, 1750–1830

This chapter explores the nature of credit and credit markets through the eyes of financially distressed and occasionally bankrupt wigmakers and coiffeurs in Sweden from the 1750s to the 1830s. The hair professionals were a specific group among artisans. They were in a bidirectional position in the credit market as they both lent and borrowed money, and the growing demand and the social and political significance of hair made hair professionals important members of early modern societies. This provided them, at least in theory, with a position where they could earn a decent living and even accumulate wealth. The chapter, however, shows how hair profession-als still were plagued by repayment problems, poverty and insolvency. This was caused not only by the ups and downs in the hair business in general but also by the nature of both their individual businesses and the credit market.

Book printing in Stockholm, from royal privilege to market economy, 1780–1850

This chapter investigates bankruptcies in the book printing industry in Stockholm between 1780 and 1850. The printers in Stockholm produced the bulk of printed material in Sweden. Books of high quality, printed us-ing new types mainly imported from foundries in France, were also made in the capital. In 1810, after a state revolution, almost all regulations re-garding the printing industry was taken away. This meant that all exist-ing printers, formerly protected by royal privileges, had to compete with anyone who wished to become a printer. The small but important book printing industry had suddenly become an experiment in market economy. This also meant that the state lost control over the industry and was forced to compete on equal terms with other political actors in an intense print-war. A large number of bankruptcies among book printers suggest that this was the first critical era of the modern printing industry in Sweden. The outcome was a small number of efficient firms, which in the late nineteenth century would form the backbone of a modern mechanized book printing industry in Sweden.

Cabinetmakers and chair makers in Stockholm, 1730–1850. Production, market and economy in a regulated economy

This chapter explores the economic conditions among furniture makers in Stockholm between 1730 and 1850, a period characterized by economic growth, global market integration, a consumer revolution and a deregula-tion of the market. Three overarching question are addressed. First, how was the production of furniture organized, how many artisans were active and what was the size of their workshops? The second question deals with the distribution of furniture. How did the artisans organize the selling of their goods and how did they reach their customers? The third question deals with their financial networks, which were essential to create and de-velop businesses. How were the workshops financed, from whom did the furniture makers borrow capital and what kind of debts did they have? The investigation is based on an analysis of bankruptcy documents from fur-niture makers, which contains information not only about their overall fi-nancial situation but also about debts, credits as well as the organization of their operations.

Credit relations among painting professionals in Stockholm, 1760–1849

This chapter explores the credit behaviour of various painting professionals, foremost including painters and wallpaper makers. Specifically it focuses on the idea about a transition to a more modern economic and social behaviour in Stockholm's luxury and fashion industry at the beginning of the modern era. The investigation explores the early modern notion of creditworthiness against name security and asks how this was replaced by comparatively higher security requirements in the wake of the reform of the bankruptcy institution. The increased proportion of loans with collateral is here interpreted as a kind of modernization compared to creditworthiness based only on general conceptions of the borrower's ability to pay based on, for example, titles, community position or property ownership. The results suggest that the arguments hardly can be described as "modern" or "rational" in a strict economic sense. In several cases, painters and wallpaper makers on the contrary made guarantees that far exceeded their assets.

Part IV: Conclusions

The Stockholm credit market in an international perspective

This chapter brings together the previous chapters in a broader analysis of the role and function of credit, fashion and luxury. It also extends the perspective by discussing if the modernization (defined as increased efficiency and executive ability) of the bankruptcy institution in Stockholm during the period of investigation was something unique for Sweden or a more general phenomenon in Europe. Particular emphasis will be placed on understanding the modernization in relation to a shift in the preindustrial understanding of the concept of credit and the emergence of short loans without security. In the transition between the early modern and the modern time-period the Swedish state took on a new role by initiating the modernization of the credit market. During this period, the bankruptcy institution was increasingly formalized and better organized to make it more operationally executive, mainly by shortening the time of the bankruptcy proceedings, but also by ensuring that cases were tried to verdict. Another example was the creation of state-operated credit institutes to alleviate cash flow problems, a major issue for producers and retailers in the fashion and cultural industries.

Notes

1 For consumption as a dynamic force see Braudel (1981); Lemire (1991); Brewer and Porter (1993); Fine and Leopold (1993); Roche (2000); Trentmann (2012); Lemire (2018). For Swedish circumstances see Ahlberger (1990); Magnusson and Nyberg (1995); Lilja (2010), p. 49. See also Lindgren (2010), pp. 95, 100–101, who

claims that promissory notes by the middle of the nineteenth century constituted more than 75 per cent of the household assets registered in probate inventories. A study of a parish in Southern Sweden in the mid-nineteenth century
has shown that promissory notes constituted between half, to two-thirds of the
local credits. Perlinge (2005), p. 76. A study of the Champagne region in France
shows that promissory notes constituted 43 per cent of the credits in 1769–1772.
Brennan (2013), p. 37.

2 de Vries (1994, 2008).
3 Finn (2003); Fontaine (2014); Muldrew (1998); Crowston (2013).
4 Söderberg et al. (1991) about Stockholm's general social and economic development. The modernization of the late eighteenth-century and nineteenth-century
   Swedish economy is discussed in the framework of a "financial revolution" in
   Ögren (2010). For general and regional European legal aspects of insolvency and
   bankruptcy, see Nadelmann (1957); Udell (1968); Ross (1974); Hansen (1998);
   Sgard (2006); Fischer (2013); MacLeod (2013); Falk and Kling (2016); Wilmowsky
   (2016).
5 Nyberg (2010); Nyberg and Jakobsson (2013, 2016); Hayen and Nyberg (2017).
6 Finn (2003), part II; Fontaine (2014), pp. 8–14.
7 This section is a revised and updated version of Nyberg (2010), pp. 19–21.
8 Mathias (2000).
9 Grassby (1995), ch. 3. In Stockholm the sponsors recorded the date on which
   a wholesaler was granted permission to trade. SSA, Handelskollegiet, AI,
   Protokoll, huvudserien.
10 Fontaine (2001, pp. 39–43, 2014, ch. 3).
11 Kermode (1998), pp. 242–247; Hunt and Murray (1999), ch. 9; Spufford (2002),
   ch. 1; Braudel (1986), p. 126; Jonker (1996).
12 Fontaine (2001, pp. 39–57, 2014, ch. 10). See also Ogilvie (2005), pp. 15–52;
   Hasselberg (1998), ch. 4.
13 Fontaine (2014), ch. 3. See also the bankruptcy database at www.tidigmoder
   nakonkurser.se.
14 The regulation of financial affairs varied greatly. For a long time, Catholic areas
   for example generally viewed the charging of interest on loans as sinful, making
   it more difficult to secure credits. Fryde (1996), pp. 107–120; Jones (1979); Hoppit
   (1987); Muldrew (1998); Agge (1934); Safley (2000).
15 The main focus of the amended legislation was a gradual formalization with
   stricter payment requirements for the debtor, including limiting the extent of
   withdrawal, i.e., the right to be released from continued claims for repayment
   following bankruptcy.
16 Göransson (1990), pp. 525–543; Ågren (1999), pp. 683–708.
17 Fontaine (2014), ch. 5; Erickson (1993), ch. 5; Göransson (1990), pp. 525–543;
   Ågren (1999, pp. 683–708, 2009, ch. 4–5); Cavallo and Warner (1999); Ogilvie
   (2003), ch. 4–5; Spence (2016); Coffin (1996); Gross et al. (1996); Pearson et al.
   (2001).
18 Nyberg (2006), pp. 155–180. Still, different approaches were discernible. Women
   with higher social status often made references to the legal arguments when arguing their case in this process. Women from lower social groups on the contrary were more likely to try and safeguard their dowries with emotionally based
   arguments (Paulsson, 2004).
19 The theory has since been expanded further and discussed extensively by researchers in social and economic history. See the most recent books by Zhan
   (2019); Hayami (2015); Palat (2015); Hutchison (2012); Muldrew (2011); Cruz
   and Mokyr (2010) and articles by Poukens (2012); Bull (2011); Ogilvie (2010);
   Saito (2010); Allen and Weisdorf (2010); Seong Ho et al. (2009); Meerkerk van

Nederveen (2008); Brown (2000); van Zanden (1999); Lee (1999); Clark and Van Der Werf (1998).

20 De Vries (2008), p. xi; McKendrick et al. (1983); Brewer and Porter (1993).

21 Zhan (2019); Hayami (2015); Hutchison (2012); Muldrew (2011).

22 Muldrew (2011).

23 Hutchison (2012).

24 The division between new luxury and old or even corrupt luxury relates to a distinction put into writing by philosophers and writers during the period. Berg (2005), p. 5. Generally on luxury and colonial wares, see Sargentson (1993); Riello and Tirthankar (2009); Riello and Parthasarathi (2009); Riello (2013); Simonton et al. (2015); Hofmeester and Grewe (2016); Hodacs (2016); Schäfer et al. (2018).

25 Berg (2005), pp. 6, 24.

26 Berg (2005), pp. 21–110; Hodacs (2016).

27 Finn (2003); Berg (2005); Fontaine (2014); Ilmakunnas (2012); Crowston (2013); Lemire (2016, 2018).

28 Berg (2005), p. 32.

29 Berg (2005), p. 330.

30 This became increasingly common in the beginning of the nineteenth century throughout Europe. The takeover of castles, farms and other production centres in the long term led to the bourgeoisie strengthening its social position at the expense of the nobility. In the end, debt was regulated based on the long-term security of the credit relationship. Although the financing for foreign trade and trade companies had a similar rationale, they eventually had to repay their operating credits. Royal lavish consumption was an extreme variant of this problem but royals could neglect their debts to a greater extent than the nobility (Fontaine, 2014).

31 In Sweden, the typical promissory notes had a maturity date of about 3–12 months with a 6 per cent interest rate. The notes could be transferred. The main forms were simple and current debentures, respectively, where the latter could be mortgaged. They could be traded indefinitely at the end of the term and could be transferred to other people. In addition to reversals, bills and assignments were common in the European credit market. See Ögren (2010) for Sweden, and for more general aspects de Roover (1948); Dehing and 'T Hart (1997).

32 See note 19.

33 Between 1777 and 1834, for example, three so-called *myntrealisationer*, or in short attempts to restore the value of banknotes in terms of a metal standard were undertaken (Ögren, 2010).

34 Nyberg (2010).

35 Muldrew (1998); Finn (2003); Berg (2005); Fontaine (2014); Crowston (2013); Lemire (2010); Campagnol (2014).

36 Roche (1996); Poni (1997), footnotes 3 and 6; Hammar and Rasmussen (2001); Jenkins (2003), part III; Richardson (2004); Kawamura (2005); Styles (2007); Rasmussen (2010); Riello and McNeil (2010); Bremer-David (2011); Campagnol (2014); Crowston (2001, 2009, 2013); Nordin (2013); Möller (2014); Tétart-Vittu et al. (2014); Nyberg (2015a); McNeil and Riello (2016); Mortier and du Bianca (2016); Cumming et al. (2017); Ribeiro (2017); Welch (2017).

37 Raj (2011), pp. 56–57, 68–69.

Part I

Paris

The capital of luxury

1 Rational follies

Fashion, luxury and credit in eighteenth-century Paris

Paula von Wachenfeldt

Introduction

THE caprices of fashion among the French are astonishing; they have forgot how they were dressed in the summer; they are even more ignorant how they shall dress this winter; but, above all, it is not to be believed how much it costs a husband to put his wife in the fashion.[1]

The astonishment of Montesquieu's fictional Persian traveller Rica in 1717 illustrates the authority of fashion in Ancien Régime France. Although this style dictate might have been confusing for an Asian foreigner at the time, it was nonetheless a social practice for the Parisian. This craving for bodily adornment required credit facilities that became crucial for the circulation of fashion and luxury in eighteenth-century Paris. The purpose of this chapter is to examine writings about fashion and credit in order to discern, on the one hand, the critique of credit practices in French society, and on the other hand, the social discourse that upheld the important consumption of fashion and luxury goods. I thus aim to explore the discursive mechanisms behind the systematic craving for luxury that made use of credit facilities. The relation between changes of style as a socio-cultural habit and monetary agreements begs the following question: *how did fashion and luxury, helped by the credit system, construct and codify Ancien Régime society?*

A great deal of research has focused on the networks that surrounded credit in society and not least the conditions of seamstresses and their bankruptcies due to unpaid debts.[2] The networks involved in the aristocratic consumption of the Parisian *hôtel*s and the important commercial activities of the marchands merciers in eighteenth-century Paris have also been subjects of investigation.[3] In addition, microstudies of debt portfolios have been used to retrace the web of creditors and the nature of the given credit.[4]

What we need to consider, however, is the history of ideas about credit, fashion and luxury that might have affected the public attitude towards consumption. In the following, I propose that credit connected to fashionable consumption is the fruit of social ideas and cultural beliefs about the deeply entrenched role of clothing in pre-revolutionary France. This postulation will be examined first in popular plays by Molière and Lesage, and second, in fashion magazines that I maintain conveyed societal values related to sartorial practices. Furthermore, Mercier's moral and physiological observations of Parisian mores and customs, known as *Tableau de Paris*, and the memoirs of the Baroness d'Oberkirch about court society constitute other important writings about French eighteenth-century society. Cultural work is an indispensable tool for understanding "the shared ways in which author and public think and discuss the world".[5] What is more, the cultural production of an era often reflects its ambitions and struggles and can therefore be crucial in the reconstruction of the social system.

It may be objected here that literacy in eighteenth-century Paris could not have been widespread, and therefore, written sources must have been a privilege of the cultural elite. In other words, ideas about fashion and luxury could not have reached the lower strata, and their magnitude was therefore restricted. Nonetheless, according to Roche, the city of Paris represented a distinct situation where, by the end of the seventeenth century, 85 per cent of men and 60 per cent of women could sign their wills. In 1789, on the eve of the Revolution, the number had increased to more than 90 per cent for men and up to 80 per cent for women.[6] Could the signing of official documents mean that the signer was able to read in general? And wasn't, after all, the making of a will a signifier of a privileged act for the better-off?

A simple signature does not necessarily entail literacy, and certainly, the act of making a will confirms a social prerogative. Nonetheless, probate inventories reveal that, around 1700, 85 per cent of men and women servants who survived their partner could sign notarial documents. Around 1789 this number was up to 98 per cent. Moreover, the servants of the city seem to have enjoyed at least a basic education that also improved over time and even signed notes of debts or account books.[7] This activity revealed by the probate inventories thus indicates a certain knowledge of the social system even in the lower strata. Mercier notes in his *Tableau de Paris* that his ancestors did not read, while in his time, almost every class could read.[8] This observation certainly does not distinguish between migrated rural (considered as ignorant) and Parisian (initiated) servants and ought therefore to remain quite general. But, all in all, based on the above-mentioned assumptions, I argue that, with a growing literacy, miscellaneous writings about credit and fashion had an impact on the public opinion. Further support for this argument is that other inventories reveal that books became more popular even among the *petit peuple*, and workers and shop assistants read nearly as much as their masters.[9]

Since I am examining the ideas of fashion and luxury, and their connection to credit, the gender perspective becomes indispensable. The *marchandes de modes* who created sartorial magic as much as they provoked social debates on their profession also illustrate well a growing view on luxury and fashion as female concerns. What is more, the retail profession that relied on the principle of credit was also the subject of critique.

Clare Haru Crowston underlines that

> [c]redit relations were universal in the garment trades. From the highest
> noblewoman to the lowliest shopgirl, seamstresses' clients almost never
> paid the full price of the garments they ordered. Instead, they made a
> small initial down payment and promised to pay the rest in instalments.[10]

The craving for luxury as an imperative activity in Ancien Régime France
entailed, on the one hand, indebted consumers, and on the other hand, the
stated bankruptcy of many seamstresses when scores could not be settled
entirely. But above all, the social feature at the heart of the credit system laid
the grounds for the idea of fashion as a commercial business and a cultural
institution during the pre-revolutionary years.

Defining fashion, luxury and credit

The *Dictionnaire de l'Académie Françoise* from 1694 highlights the idea of the
limited duration of fashion and the human whim that makes it changeable.[11]
The *Encyclopédie* from 1765 stresses the features of ornaments for both sexes
that rely on the elements of change and caprice. The view of fashion as a
sensitive and foolish trend in society is hence far from being a contemporary
view.[12] As to the notion of luxury, its definition from 1694 embraces the aspect of lavishness including the features of clothing, furniture and tables.[13]
Later in the *Encyclopédie*, Saint-Lambert describes it from different angles,
oscillating between the critique led by the moralists against it and its benefits for society. The article demonstrates further the inconsistency in the
different presumptions about luxury which results in a correctly nuanced
and complex picture.[14]

I have argued in an earlier study that the concepts of fashion and luxury
should not be mixed in contemporary society as the first one relates to
trend-sensitive and mass-produced objects, while the second implies qualified design and a long experience of craftsmanship.[15] This distinction seems
to have appeared in the eighteenth century as, in the definitions above, the
notion of fashion is based on caprice, whereas luxury focuses on the extravagance of certain material objects. In other words, fashion is change while
luxury is superfluous. That said, the metamorphosis of style and fashion in
pre-revolutionary France was in reality the luxury of the better-off aristocracy and the middle class. Therefore, both concepts need to be dealt with in
this study, but we will also see the transformation of the view of luxury and
the new image it embodies due to the impact of fashion.

The meaning of credit that arguably supported the consumption of fashion and luxury in Ancien Régime France is also interesting to bring to light.
Le Dictionnaire de l'Académie Françoise from 1694 defines credit as based
on the good reputation of the borrower that thus allows him to borrow.[16]
The features of morality, commerce and personal and real (economic)

security appear in The *Encyclopédie* in 1754.[17] The idea of a good reputation and the ability to meet obligations are central and repeated in 1771 in the *Dictionnaire universel françois et latin* where credit is defined as a mutual loan based on money and commodities built on the reputation, honesty and solvency of the merchant.[18]

All of these definitions underline the common traits of the trust, virtue and reputation of the buyer. Credit thus relied on the immaterial side of human beings as much as the material affair itself. But as credit inevitably generated both commercial and human relationships with the client, merchants could never really clear an account.[19] Credit hence remained a frequent social activity in the trade world of early modern society.

Fashion, luxury and credit in writings

In her memoirs about the court of Louis XVI (r. 1774–1792) and pre-revolutionary French society, the Baroness d'Oberkirch tells us that "bankruptcies were everywhere" and that the empire of fashion endured a big collapse when the "insolent" seamstress of Queen Marie Antoinette, Rose Bertin, went bankrupt.[20] In the course of the eighteenth century, debts were a recurrent feature, even in lower-class society. The nature of debts could vary from rent to charges of medical care and burial with an increase of the last two over the course of the century. Interestingly, in our context, debts for groceries, chocolate, meat and baked goods doubled. This indicates that, besides the necessary need for health treatments, poor servants and workers got into debt to enjoy a better standard of life.[21] Furthermore, clothing acquired a great importance in eighteenth-century Paris due to, if not fashion, then the "ambience of Parisian life" that encouraged wearing the new.[22] By examining the writings of the time, this vague state of "ambience" as described by Daniel Roche can become more concrete, giving evidence of the ideas and beliefs that circulated in society. Hence, the question we need to ask ourselves is: how did the commerce of fashion flourish and become a social desideratum? What cultural channels might have affected the public opinion and thus created an urge for consumption of clothing that, in consequence, entailed the need for credit?

The theatre

Presented for the first time in 1668 at the Palais-Royal, *L'avare* [*The Miser*] by Molière received a frosty reception due mostly to its prose form, which was unusual at the time. The intrigue of love in the play is disrupted by other motifs as money obsession, shady credit practices and unscrupulous behaviour of usurers. Cléante, the son of the greedy Harpagon, needs to borrow 15,000 francs and engages his valet La Flèche to arrange the affair. The lender will lend his money at no more than five-and-a-half per cent, which at first appears reasonable. His conditions are quite outrageous, however. As he does not have the money himself, he needs to borrow it from another broker at the rate of 20 per cent. In addition, as the lender has only 12,000 livres in cash, the borrower will have to take chattels, clothing and jewels from the lender, the value of which ought to correspond to the remaining 3,000.[23] At the heart of Molière's comicality resides a social satire of greed that leads to an abusive credit system and exploits the needs of the borrower. Frosine, the intriguing woman in the play, is also in need of money, and despite her effort to serve and please the master Harpagon, he remains deaf to her request. In reality, servants in Ancien Régime France were paid both in kind and in specie.[24] Frosine, as a servant, is requesting payment in specie as an exchange for her services to her master. There is a mockery here of the dubious model of payment that is at stake when greed becomes predominant. Cléante, the son, summates the main critique of the play: "Who is the more criminal in your opinion: he who buys the money of which he stands in need, or he who obtains, by unfair means, money for which he has no use?".[25] While the borrower's needs, based on a good cause according to Cléante, are defended, the dealer is criticized for exploiting the credit system.

The play also addresses the traditional social critique of clothing and the costs connected to it. The miser Harpagon condemns the lifestyle of his son that allows extravagant charges for ribbons and wigs. Cléante admits that he plays and all profit is spent on clothing.[26] The meaning of dress during the reign of Louis XIV (1643–1715) onwards is linked to the values of distinction and extravagance. Sartorial signs contained a complex system of sensibility and an ambition for social mobility, leading to its importance for the social body. The fact that Cléante spends all his profit on adornment shows not only the influence of fashion on French society but also a bourgeois gentleman's pursuit of refinement and adherence to a noble appearance. The system of credit was certainly a breeding ground for abuse but it opened up two possibilities: on the one hand, it helped women workers to enter the fashion market as was the case for the *marchandes de modes*, as we shall soon see, and on the other hand, it enabled social mobility in a society with visually recognizable ranks.

Another play from Molière highlights the role of appearance in the Ancien Régime. Monsieur Jourdain in Molière's *Le bourgeois gentilhomme* from 1670 illustrates the desperate attempts of a bourgeois man to imitate the distinguished aristocratic world. In the second scene, Monsieur Jourdain explains to his music and dance teachers that he is late because he had to dress for the occasion like "les gens de qualité". He emphasizes further that his tailor has sent him stockings in silk, which was a great luxury at the time. Monsieur Jourdain is also keen on showing his new livery to the present teachers and he engages different people to learn how to play music, dance, converse, fight and bow to a marquise.[27] Molière often satirized the anxiety of the middle class in its quest for noble manners and appearance. As a way to denounce social practices, Molière used the communicative nature of clothing. The style, the cut and the accessories thus became a medium for depicting fashion victims and their derisory and excessive behaviour.[28]

Going back to Harpagon's disapproval of sartorial ornament, one can distinguish the harsh discourse that accompanied the view of luxury and its consumption throughout history.[29] The condemnation of bodily adornment is based on its irrational nature. "If you are lucky at play, you should profit by it, and place the money you win at decent interest, so that you may find it again some day" says Harpagon to his son. While money placement can generate interest and growth – in other words, investments and security over time – expenses for fashion items engender unreasonable actions since fashion, by definition, is frivolous. In fact, sartorial commodities create a hierarchy of values as they are considered a luxury. Consequently, the consumption of these goods expresses gratuitous and unjustified behaviour. Rousseau's severe attack on luxury in 1754 and 1755 can be seen as part of a growing criticism of a materialized world. For him, the consumption of commodities generates exchange values and, not least, harmful effects on human behaviour.[30] Although Rousseau did not specifically condemn the issue of credit, it can be argued that the economic facilities it entailed promoted the consumption of fashionable follies that Molière's characters are blamed for.

The play *Turcaret* from 1709 by Alain-René Lesage is the story of the trickery and dishonesty of a financier, a knight and a baroness who cheat each other financially in order to achieve their own goals. The intrigue is a denunciation of immorality caused by the desire for money. Turcaret is a parvenu who has become a so-called "traitant" or a broker who collects taxes and credit for the king but who abuses the system in order to enrich himself. He has fallen in love with the baroness, overwhelming her with gifts and making her believe that he is a widower. The knight cannot pay his debts and therefore pursues the baroness in order to ruin Turcaret. The only reasonable and ingenious people in the play appear to be the servants who are well aware of the fraud in which all the characters are involved with. Frontin, the knight's valet, is the best one to resume the series of falseness in this intrigue: "I admire the train of human life! We pluck a coquette, the coquette eats a businessman plunder others: this is the most pleasant ricochet of deceit".[31]

In the manner of *The Miser*, *Turcaret* illustrates the avidity of the brokers and their scrupulous business. Much more than the former play, it is almost utterly centred (with no significant love intrigue) on the credit trade and the dishonesty connected to it during the last years of the reign of Louis XIV. *Turcaret* illustrates the dissipation of moral values caused by a social system that creates needs while supporting greedy financiers. It is thus less surprising that the financiers of the time tried to stop the play at *La Comédie-Française*, blaming it all on the cold weather. *Turcaret* stayed on the repertoire, however, and became one of eighteenth-century's most successful plays.

Fashion periodicals

The printed press expanded substantially in the eighteenth century, even reaching provincial towns and the colonies. In his study of literature and the press in seventeenth and eighteenth-century France, Jurt Joseph highlights that this kind of printed matter was often the only cultural institution for members of the lower social classes.[32] His assumption supports the earlier argument about the increased literacy of the popular classes and Mercier's own observation of widespread reading in the eighteenth century. Thus, fashion periodicals, as a popular press, were arguably one of the most important vehicles of taste dissemination at the time both in Paris and the provinces.

The *Gallerie des modes et costumes français* was the first magazine launched in 1778 to be completely dedicated to the latest fashion. The text of this first volume is informative as every dress, hairstyle and hat are meticulously described and accompanied by images and the addresses of the shops. In 1785, *Gallerie des modes* was replaced by *Cabinet des modes ou les modes nouvelles*, which in turn transformed, a year later, into *Le Magasin des modes nouvelles, françaises et anglaises*, lasting until the French Revolution of 1789.[33]

The first volume of *The Cabinet des modes* from 1 January 1785 informs the reader about the significance and utility of luxury in society. The magazine claims, in accordance with the contemporary Voltairian view, that luxury advances the commerce, helps the poor through its production and can therefore restore inequalities in society.[34] It further advances that the consumption of luxury can encourage manufacturing, the creation of the Beaux-Arts and the success of agriculture. The making of luxury, as it is explained in the magazine, entails more resources and greater pleasures.[35]

Even though this volume discusses the consumption of luxury in particular, the consumption of ordinary fashion was a reality even for the lower classes. Maid and manservants came in contact with this kind of consumption not only through their masters but also in Taverns, theatres and at fetes and fairs, where sartorial taste could be passed on.[36] Cultural festivities, like the theatre, might have played an important role in the dissemination of fashion which can be seen in fashion periodicals. *Le magasin des modes nouvelles, françaises et anglaises* from November 1786 presents for example the hats à la Randan that owe their taste to the actress Mademoiselle Contat in the play *Les amours de Bayard* by M. Monvel. The details of the dress, accessories and hair are meticulously described, enabling the reader to grasp every part of the look. The theatre seems to be an inspiring space in fashion matters as the same magazine reported on how an elegant lady had appeared lately dressed in a multicoloured fringed muslin scarf, tied in the front with a long gold pin.[37] What is more, all the depicted garments can be purchased at "le sieur Jubin".[38] The space of theatre and opera continued to be inspirational as even the volume of December 1786 reports on ladies who were seen wearing different styles of hats. Interestingly, the popularity of this magazine appears through the report on provincial subscribers who did not receive it.[39] This joins the initial argument about the spread of periodicals even in the provinces, which in turn helped the circulation of fashion.

In the volume of 20 December 1786, the editor of *Le magasin des modes nouvelles, françaises et anglaises* described the aim of the magazine as spreading French and English fashion to the whole universe and satisfying people's passion for the commodities that appear with greater advantage and splendour through *la Mode*. Another important objective was to serve the traders and manufacturers by instructing them about the most desirable fabrics and colours, the design of furniture, jewellery, carriages and plates.[40]

The statement suggests how the press was attributing itself with substantial power on taste and lifestyle and therefore on commerce. This power was even extended to the dominating role that *Le magasin des modes nouvelles* considers itself to have in fashion matters even on an international level. This impact had however been interrupted by recent claims made by German, Saxon and Belgian subscribers about the poor quality of the magazine. The investigation shows that the librarian Tuttot in Liège was the author of a forgery of the magazine. The upset tone can be connected to a perceived threat to the French fashion hegemony. This forgery was seen as harmful to the traders and manufacturers since they were usually well informed and directed "with confidence" by the magazine.[41] The argument of this worriment involves the idea of trust accorded to the magazine as a leading expert in fashion matters and, therefore, as an essential element of support for the commerce.

In line with the growth of consumption of fashion in society, the role of the magazine was hence to advance a discursive practice about its priority for French society.[42] Meanwhile, the press can consolidate its important mission as a tastemaker helping both commerce and consumers. Indeed, the power of fashion periodicals is not to be disdained as recommendations on the latest must haves are made very clear in *Le magasin des modes nouvelles*. Examples include the *bonnets de nuit* that women wore under their *chapeau-bonnette* when they went out in the morning. This fashion was condemned by the magazine as not many women would look appealing in it.[43]

In the volume of 10 January 1787 the author admits that French ladies, during the previous two years, have been borrowing foreign styles from Poland, England, Turkey and China. Nonetheless, the French lady could put her own touch on these styles and embellish them, making them even better than the original: "Quand elles copient, elles corrigent, elles embellissent".[44] The tone here is patriotic, as French fashion appears to have a high-cultural value for the state.

These periodicals were also an important medium for the dissemination of Parisian taste and style in the provinces. Madame d'Oberkirch reveals in her memoirs that, despite the fact that they were in the countryside, they were still well informed about Parisian fashion and news in 1787 and that these gazettes were sent to them by friends. Reporting on the latest news in this same context, she explains that luxury and wealth in the form of beautiful fabrics and jewellery were still dominant in the capital.[45]

If *Le magasin des modes nouvelles, françaises et anglaises* reported on both French and English fashion, as indicated in the title, another periodical from 1799 proscribed the impact of English fashion on French taste and the importing of foreign goods. In the first volume of *Tableau général du goût, des modes et costumes de Paris*, it was made clear that fashion only knows one power, which is the public spirit. The importance of luxury for manufacturing, the French trade and, consequently, the national growth is also highlighted. The tone is quite firm, on the one side blaming English fashion influences (ladies ought to buy French gloves instead of English ones) and the irresponsible driving of carriages and, on the other side praising domestic taste and the manufacturing of luxury.[46] Fashion, luxury and behaviour are subjects taught by the French press. We can also see how the advent of fashion magazines in the eighteenth century enabled the mediation of fashion to the people, a trend that persisted up to modern times.

Another important communicative tool was advertising, examples of which can be found in the weekly magazine *L'Avantcoureur*, which contains miscellaneous short articles about, among other things, plays, music, geography, medicine and architecture. The volume of August 1763 contains information about the point of purchases and price of two Chinese figures dressed in their country's costume. In addition, brands for cosmetics and pomade for the hair made from the fat of the bear's neck are announced with prices and address.[47] The advertisement for hair appears again in the volume of 12 February 1770, where the reader can be helped with hair colour that does not damage the skin and prevents hair loss.[48] Furthermore, almanacs were also used for shop advertisements and the diffusion of fashion plates.[49] We see here how commerce started to rely on the power of advertisement as a way of attracting a broader range of consumers. The role of advertisement is to build a new legitimacy in fashion matters that is able to shape demand. The link between periodicals, advertisements and demands has been substantial when developing novel tastes, needs and styles in society.

Luxury on credit: mercers and marchandes de modes

The establishment of shopkeepers in Paris in the second half of the eighteenth century contributed greatly to new forms of consumption, in particular of exotic products.[50] Rue St Honoré and Rue St Martin were considered as the capital's "grandes rues de commerces", where mercers (merciers in French) who were specialized in the sale of fashionable clothing, furniture, fabrics, jewellery and hardware were located. Mercers were men, and their businesses were involved with manufacturers and suppliers, and they even designed their own products. Nearly half of them had shops in Rue

St Honoré in 1769.[51] Mercers had the status of merchants, and they considered themselves as unique since they did not do manual labour. The earlier image of a mercer from the late seventeenth century as a poorly dressed street peddler was outdated by the middle of the eighteenth century as these were now dressed-up shopkeepers.[52] First established in Les Halles in the seventeenth century, the mercers followed the luxury trades and set up their boutiques on Rue Saint-Honoré, Rue Saint-Denis, the quartier Saint-Eustache and near the Palais-Royal in the eighteenth century.[53]

One of the most distinguished mercer's shops at the time was *Le petit Dunkerque*, founded in 1770 by M. Granchez and situated at the Quai de Conti.[54] Queen Marie Antoinette was a regular customer and even Voltaire seems to have been a visitor, at least if we believe Mercier's article about the shop. Mercier described it as a jewellery boutique near Pont Neuf where one could find all of the frivolous jewels given to the "femmes honnêtes". These ladies, according to Mercier, would not accept money but rather trinkets in gold as those looked better. He maintains further that the jewellery purchased by the lords was on credit and distributed to ladies with an air of nonchalance. Such expenses exceeded the necessary and were therefore considered as luxury.[55]

Purchasing jewellery on credit was common in the Ancien Régime, and in this context, three kinds of credit can be distinguished: the short, the medium and the long term. In Paris, short-term loans were typical from merchants or bankers to other merchants, usually to be repaid within three months or less. Unlike long-term loans, the short ones were often unregistered private agreements.[56] As mentioned earlier, credit entailed the important relationship with the client and a nobleman or respected burgher who entered *Le petit Dunkerque* was likely accorded medium or long-term credit for the piece of jewellery he was buying. Although Mercier started by mocking the lavishness of such expenses and the credit they involved, he acknowledged the skill of the master Granchez. According to him, Granchez was able to imagine and initiate a new taste in society that, in consequence, would make it possible to employ people from the capital instead of bringing them from abroad at a great expense. On a production level and from a worker's point of view, luxury was viewed as a positive force in society. Paradoxically though, its consumption was regarded as irrational.

It is against this backdrop of demand for the latest luxuries that the *marchandes de modes* became a notable female phenomenon. Although both women and men could work as *marchandes de modes* (Jean-Joseph Beaulard for example was a man), it was mainly women who had this occupation by the second half of the eighteenth century.[57] They were not seamstresses as they did not sew and cut the garments in a workshop. In reality, they had a much more decisive role in establishing new trends in society and, above all, they had a much higher social status than seamstresses. The term "marchandes de modes" appears in the *Encyclopedie* in 1765 where it was described as sellers of "ornaments and amenities" for men and women, their role being to create *la mode* of the season. They also did hairstyles and put them up like the hairdressers. Having originally been part of the *mercerie* commerce, they had left that trade for fashion.[58]

The marchandes de modes added decorative touches to garments and hairstyles using artificial flowers, feathers, ribbons, lace, gauze and linen. As Jones put it:

> In theory the *marchandes de modes*, like the mercers, sold and embellished rather than produced goods; in fact the *marchandes de modes* made several items of apparel, including outerwear cloaks and capes and a variety of scarves, shawls, and mantles.[59]

Marie Antoinette's favourite *marchande de modes*, Mademoiselle Bertin with her shop *Au Grand Mogol*, gained excessive power and a solid reputation in many countries.[60] Baroness d'Oberkirch deemed her as imperious and "puffed up with her own importance".[61] Despite this dislike, the baroness still admitted her success and wonders and the fact that everyone wanted to tear off her hats.[62] Bertin was not just a *marchande* but became friends with Queen Marie Antoinette and visited her twice a week to order her wardrobe. The business relation was, like in society as a whole, built on credit, as the Queen rarely paid for her purchases.

The Baroness d'Oberkirch wrote again in 1787 that the empire of fashion had suffered a great cataclysm as the "insolent" Mademoiselle Bertin had gone bankrupt – and this was not a plebeian bankruptcy, but one that amounted to about two million livres![63] This alleged bankruptcy has not been confirmed and seems to have been a fabrication to obtain overdue payments from the "trésor royal" and shame the crown.[64] The satirical short text *Grande dispute entre Marie-Antoinette et ses fournisseurs, traiteurs, tailleurs, marchandes de mode, &c., &c., &c.,* from 1792 by Nicolas Prévost, illustrates the Queen's large debts. The sole scene in the text between a food supplier and Marie Antoinette shows her mindless attitude and her inability to understand the former's need to be paid. Instead, the Queen suggests that he should accept credit again. Though she feels obliged to pay the food supplier 200 livres, she ends up regretting it and snatching his wig after hearing the acerbic judgement: "You have long deceived France". This common critique of the Queen – her refusal to follow court etiquette and her bad political influence on the King – overshadowed her ostentatious lifestyle in a heavily mortgaged France.

Between 1771 and 1788, the budget of the Queen's household (divided between fixed fees and unpredictable costs) increased from 1,056,000 livres to 4,700,000 livres. The most expensive year of her reign was 1785, when the wardrobe budget showed costs of 258,002 livres (compared with a tailor's monthly salary of 4 livres at the time). Interestingly, the costs of fabrics, amounting to 65,929 livres, decreased in 1780 in favour of costs associated with fashion items which reached 120,284 livres.[65] It was the urge for newness – and not excessive fabrics – that caused unreasonable luxury expenses, all ordered and enjoyed through credit. The idea of newness and short-lived trends thus became central in the formation of the eighteenth-century fashion market.

We have already seen the criticism of the shady practices of credit addressed by Molière. The consumption of clothing, considered as a luxury and therefore irrational, embodied another disruptive subject. The eighteenth century follows up this criticism through the discourse held against luxury which, in Rousseau's words "impoverishes everyone else, and depopulates the state sooner or later".[66] The Physiocrats were harsh attackers of mercantilism and manufacturers, suggesting that wealth could only come from the land.

Nonetheless, there was a new authoritative female figure in society that helped to speed up the consumption of luxury. This was the *marchande de modes*, who came to control the taste and sartorial value – in other words, both the culture and the economy of fashion. This player on the market caused moral worries for contemporaries. A request dated 1790 addressed to the mayor of Paris with the title *Étrennes aux grisettes* argued against the "marchandes de mode, Couturières, Lingères and other commercial grisettes on the pavement of Paris", presenting a harsh accusation of prostitution and boundless commercial activities. Interestingly, these different categories of female workers are bundled altogether under the same label, namely the *grisettes*, who were Parisian young women working as seamstresses, milliners, flower-makers or shopkeepers, and stereotyped as having loose morals. As Jones puts it, from the second half of the eighteenth century we can see a "feminization" of the fashion trades: "In conjunction with the movement of large numbers of women into the fashion trades, contemporaries came to believe that the production of clothing was particularly appropriate work for women".[67] The question we need to address at this point is around the difference between luxury as an exclusive and durable form of craftsmanship and fashion as a trend-orienteed and changing sartorial phenomenon.

Luxury has existed since ancient times as a crucial expression of higher rank, considered as the lord's duty towards his own society. Embodied in art, palaces and furniture, luxury was the representation of the political and social power of monarchs and the nobility. The quality of fabrics, as a consequence, was in the service of this same socio-political authority. Nonetheless, from the second half of the eighteenth century it would no longer be the quality of the fabrics that would be important but rather all the accessories that embellished the garments.

I would suggest that the role of the *marchandes de modes* can be attributed to the transformation of conventional luxury (old luxury) to the fashionable and changeable one (new luxury). Fashion, in the hands of these merchants, became the *new luxury*, namely an institutionalized capitalistic system based on the frequent renewal in style. This can be illustrated with the portrait of Marie Antoinette by Vigée Le Brun that caused a scandal when it was displayed at L'Académie Royale in 1783. The Queen was painted in a *chemise* made of muslin wearing a hat of straw with no apparent jewellery. The critics could not accept having the Queen of France portrayed in a simple dress made from cotton. Moreover, the transparency of the fabric entailed an intimacy that was inappropriate for a Queen. Vigée Le Brun therefore had to rework it with the new title "Marie-Antoinette à la Rose" instead of the first one, "Marie-Antoinette en chemise".

The crux of the matter lies in how this new fashion style composed of "simple" fabrics was deemed unsuitable for the higher social strata. What the critics did not want to accept was the changing form of luxury – it was not necessarily the fabrics that had to be luxurious but the way in which the dress could be styled and arranged. The *marchandes de modes* were the authors of this transformation of the view of luxury from a rigid and steady expression in style and construction to a capricious and playful one. This is fashion as we know it today. More important still, the *marchandes de modes*, through this process, established the idea of fashion as a commercial business relying largely on credit. In fact, the establishment of these so-called "follies" would, in all likelihood, not have been possible without the system of credit that made their existence possible. Through the medium of credit, female merchants were able to create fashion as an economic and cultural institution – a foolish, yet rational and imperative business.

Conclusion

When depicting bankruptcies, Mercier states that the cause of their common presence is due to the merchants who have left behind the old modesty of their social status. They knew luxury and opulence, and they took on a completely different appearance than the one dictated by their profession.[68] We see here two major issues connected to the consumption of luxury – the one that leads to debts and insolvency and another that makes people lose their virtue.

The aim of this chapter was to understand the history of ideas behind the establishment of fashion and luxury as a social practice in pre-revolutionary France and its reliance on credit. The examination of contemporary writings and popular plays unveiled a critique against the shady credit practices, whereas fashion magazines upheld a social discourse for the promotion of fashion consumption. A major argument that comes out of this is that the establishment of fashion as a business in the second half of the eighteenth century was essentially due to credit facilities. Certainly, the impact of the *marchandes de modes* as tastemakers is indisputable, but this impact was also the fruit of aristocratic power as it involved social networks and credit. The circulation of bodily adornment in eighteenth-century Paris was a social and cultural constraint as much as a business-driven institution. Finally, fashion and luxury not only codified Ancien Régime France but also laid the grounds for a continuous codification of contemporary societies.

Notes

1 Montesquieu (1777), vol. 3, letter XCIX.
2 See Crowston (2001, 2013); Fontaine (2008/2014).
3 See Coquery (1998); Sargentson (1996).
4 See Fontaine (2008/2014).
5 Fontaine (2008), p. 5.
6 Roche (1987), p. 199.
7 Roche (1987), p. 200.
8 Mercier (1782–1788), vol. 9, p. 334.
9 Roche (1987), p. 214.
10 Crowston (2001), p. 164.
11 *Dictionnaire de l'Académie Françoise* (1694), vol. II, p. 75.
12 *L'Encyclopédie* (1765), vol. X, p. 598.
13 *Dictionnaire de l'Académie Françoise* (1694), vol. I, p. 672.
14 *L'Encyclopédie* (1765), vol. IX, p. 764.
15 von Wachenfeldt (2018).
16 Le *Dictionnaire de l'Académie Françoise* (1694), vol. I, p. 289.
17 L'Encyclopédie (1754), vol. IV, p. 445.
18 *Dictionnaire universel françois et latin* (1771), vol. III, p. 4.
19 Fontaine (2014), p. 250.
20 Burkard (2010), p. 653.
21 Roche (1987), p. 84.
22 Roche (1987), p. 88.
23 Molière (2009), scene I, act II.
24 Roche (1987), p. 86.
25 Molière (2009), scene III, act II.
26 Molière (2009), scene V, act I.
27 Molière (1963), scene II, act I.
28 See, for example, the harsh mockery of fashion in Molière (1682a), scene I, act I and (1682b), scene I, act II.
29 Ribeiro (2003).
30 See von Wachenfeldt (2013).
31 Lesage (1709), scene XIII, act I. My translation. The original text is "J'admire le train de la vie humaine! Nous plumons une coquette, la coquette mange un homme d'affaires en pille d'atures: cela fait un ricochet de fourberies le plus plaisant du monde".
32 Joseph (2013), p. 88.
33 Although *Le cabinet des modes ou les modes nouvelles* was the "real" first fashion magazine, the *Journal de Paris* that was launched in 1777 also had references to fashion. The volume of 20 February 1777 had for example four plates of two different hairstyles with a short notice about the "Modes". The editor explains that fashion is not always easy to explain and thus it is preferable to rely on the images.
34 See von Wachenfeldt (2013) about the views on luxury in eighteenth-century France.
35 *Le Cabinet des modes ou les modes nouvelles* (1785), vol. 1, pp. 4–6.
36 Roche (2000), p. 207.
37 *Le magasin des modes nouvelles, françaises et anglaises* (1786), vol. 1, pp. 2–5.
38 Even volumes from 30 November 1786 to 30 December 1786 contain advertisements for "le sieur Donnet", the milliner.
39 The volume of 20 December 1786.
40 *Le magasin des modes nouvelles* (1786), vol. 4, pp. 25–27.
41 *Le magasin des modes nouvelles* (1786), vol. 4, pp. 27–29.

42 The field of consumption in the eighteenth century has also been subject of investigation but it is noteworthy to mention in this context that inventories from 1789 could show an increase in accumulated expenditure where linen and clothing accounted for 80 per cent of the growth. "The increase in consumer goods was general, but everything that related to the expression of appearances, both social and private, increases still more". Roche (2000), p. 213.

43 *Le magasin des modes nouvelles* (1786), vol. 5, p. 36.

44 *Magasin des modes nouvelles, françaises et anglaises* (1787), vol. 6, p. 41.

45 Burkard (2010), p. 652.

46 *Tableau général du goût, des modes et costumes de Paris* (1799), vol. 1, pp. 9–12.

47 L'Avantcoureur (1763), vol. 31, pp. 489–492.

48 L'Avantcoureur (1770), vol. 7, p. 101.

49 Delpierre (1996), p. 178.

50 Coquery (2009), p. 121. For more in-depth information about Parisian shops, see Coquery (2006).

51 Sargentson (1996), pp. 18, 44, 52, 54.

52 Sargentson (1996), pp. 9–10.

53 Roche (1996), p. 277.

54 There is an advertisement of Le petit Dunkerque in *L'Avantcoureur* about more advanced "bassinoires" or warming pans that can be used during travel, at the theatre or to warm mattresses (1770), vol. 50, p. 789.

55 *Tableau de Paris*, vol. 7, p. 82.

56 Hoffman and Rosenthal (2000), pp. 7–8.

57 Jones (2004), p. 95.

58 *L'Encyclopédie* (1765), vol. 10, p. 598.

59 Jones (2004), p. 92.

60 Other marchandes de modes that served Marie Antoinette were Madame Éloffe, Jean-Joseph Beaulard, Madame Pompée and Madame Noël. It was, however, Bertin who had the most intimate relationship with the Queen.

61 Burkard (2010), p. 196.

62 Burkard (2010), p. 424.

63 Burkard (2010), p. 653.

64 Delpierre (1996), p. 167; Crowston (2009), pp. 192–194.

65 Most of Marie Antoinette's wardrobe has been destroyed. There is, however, an inventory of her dresses in 1782 at the Archives Nationales (Series AE I 6 no 2) named *Gazette des atours de Marie-Antoinette. Garde-robe des atours de la reine*, which provides us with this information. A facsimile of this inventory has been published by the Réunion des musées nationaux, Archives nationales, pp. 9, 16, 17.

66 Rousseau (1992), p. 78.

67 Jones (2004), p. 97.

68 *Tableau de Paris* (1783), vol. II, p. 44.

2   The French model and the rise of Swedish fashion, 1800–1840

Klas Nyberg

Introduction

To what extent did France contribute to the early rise, growth and spread of fashion in Europe in the seventeenth and eighteenth century? While sumptuary laws as well as research on the leading European courts as cultural disseminators have been traditional points of departure for all types of research about the origins of fashion, the development in the receiving countries has been neglected and little explored. This includes aspects such as the emergence of domestic fashion magazines, the importance of handicraft and domestic luxury and textile manufacturers as creators of fashion. Recent research on European peripheries suggests that state-sponsored manufacturers helped to spread fashion in a more active, but at the same time locally adapted way than if the fashion models had been imported directly from their European points of origin or had been exclusively governed by sumptuary laws.[1]

Fashion as a concept is often defined as the spirit of the time with associated ideals, primarily expressed through fabrics, garments and other accessories. Fashion in the early modern period was as mentioned primarily regulated in sumptuary laws – state-defined ordinances that regulated the dress codes for various social classes – that both prescribed an allowed dress code and banned certain types of garments and materials. With the rise of industrialism, in parallel with the emergence of modern society after the French Revolution, the concept of fashion was gradually spread throughout society to affect and include all citizens.[2]

State-supported manufacturers were directly and indirectly managed by various institutions that regulated the production of consumption goods next to the production performed by the guild-regulated handicraft industry. The production in the manufactures both included high-end textiles (silk, cotton/linen), luxury items (furniture, instruments, lacquer work) as well as hedonic goods (sugar, tobacco), but also included the oldest forms of organized, standardized mass production, such as woollen fabrics for uniforms and the manufacture of weapons. Around this production – where Colbert's French system is the most well-known – an extensive legislative framework was constructed. The institutions included state departments, credit institutions and banks as well as various offices and courts to regulate and oversee work and to control and inspect the quality of the manufactured goods.[3]

The specific design of the frameworks that regulated manufacturing varied widely between different European countries and principalities, but often included prohibitions on the import of identical competing goods, detailed regulations for imitation products that were standardized models as well as royal, princely or governmental privileges to producers often in the form of monopolies.

A constant endeavour was to achieve the highest possible qualities for products such as silk fabrics, exclusive furniture, faience and porcelain. While sumptuary laws were normative statements whose compliance remain uncertain and may be questioned, the impact of European manufacturing on fashion and the growth of the consumer goods industry can be substantiated in much more detail. At times, well-preserved source materials after the government agencies and institutions that regulated and promoted the emergence of key elements of the fashion process can be consulted. In Sweden this includes records of domestic sheep breeding for the production of quality wool, silk cultivation, yarn spinning, as well as information about a number of support schemes, and subsidies as well as loans at subsidized interest rates.[4]

This chapter will show how French influences had a large and wide impact on fashion in Swedish early nineteenth-century society. I will focus on textiles, fashion magazines and the emergence of advertising 1800−1840, but the French impact also applied to architecture and interior design as well as furniture production. Overall the whole upper class culture was inspired by France, perhaps best epitomized in the use of the French language both for writing and speaking. Aristocratic families, architects, furniture artisans, silk manufacturers, designers, cultural intermediaries and, not least, the Court and the king all had French aesthetics as their prime model.

A painting of a bourgeois salon

On 30 March 1811 a proclamation for the upcoming marriage between the Court Chamberlain, Baron Gustaf Ridderstolpe and his bride, Jeanette Müller the foster daughter of the merchant and trade councillor Jürgen Christoffer Müller was read out in Storkyrkan, the main church in Stockholm, located in the Old Town, next to the Royal Palace. The wedding on April 26, that same spring, took place in the salon, the most spectacular room in the residence owned by the bride's father. The house was located at Skeppsbron, no. 2, on the south-eastern side of the Old Town, at one of the most prominent addresses in Stockholm.[5]

At the time the salon had five windows, was furnished with silk-covered furniture and gilded tables, and decorated with marble statues, ornaments and luxury items.[6] Its appearance has been preserved in a painting by Pehr

Hilleström entitled "Vid morgontéet" depicting the taking of the morning tea (see Figure 2.1). A later description of the imagery in the painting reads:

> *An elegant salon with walls and furniture in yellow as well as doors, window niches and panels in grey at the residence of the trade councillor Jürgen Christoffer Müller, at Skeppsbron no. 2. Six people are gathered around the table, including the trade councillor himself, dressed in a brown nightgown and hat, talking to the vice president Mattias Haak, who sits with his back to the observer. Three ladies are seated on the opposite side of the table, one of them reading the newspaper; behind them stands a gentleman drinking tea. In front of them the small foster daughter Jeanette can be seen standing. A maid is standing on the left hand side in the foreground by the fire pot, pouring water into a tea pot. A precious Persian rug lies on the floor, a sculptured table with writing paraphernalia, and a pair of elegant chairs can be seen on the right. In the background three antique statues can be spotted.*[7]

According to the probate inventory drawn up after the death of Müller in 1831 the residence consisted of a cabinet, the bed chamber and antechamber previously used by himself, a separate similar suite used by his wife, a salon, a hall, as well as several other rooms and chambers for the domestic staff. The stately salon at this point was furnished with the following pieces:

> *A sofa and 18 mahogany chairs covered with silk fabric, 2 gilded tables with stone slabs, 2 tea tables, 5 sewing tables, 1 mirror, 19 framed oil*

Figure 2.1 Vid morgontéet.

paintings, 2 chandeliers, 2 urns made from porphyry and a further 2 made from alabaster, 2 marble busts, 1 light piece on a pedestal, 6 bronze candlesticks, 1 glass tray, 2 decanters and 3 glasses, 1 tablecloth, 5 pairs of window curtains, 5 roller blinds, 1 floor mat and a sandbox.[8]

The marriage between the merchant's daughter and the baron saw the joining together of what has been referred to as new luxury, here meaning a socially broadened luxury for enjoyment and pleasure for a wider middle class, with the old luxury of the aristocracy.[9]

Jürgen Christoffer Müller was one of the members of a group of wealthy merchants in Stockholm with houses along *Skeppsbron*, the easternmost part of the Old Town directly overlooking Stockholm harbour. The group consisted for the most part of wholesale merchants, often of foreign origin, who in a short term had acquired great wealth, in many cases from small means, especially after engaging in the export of bar iron, which Sweden was a world leading producer of until the end of the eighteenth century.

Their residences, along Skeppsbron, were manifestation of a new kind of luxury also in the geographical space. The many seventeenth-century aristocratic palaces that had been erected in the centre of the capital had lost their importance by the end of the eighteenth century. Skeppsbron instead emerged as one of Stockholm's foremost streets with a uniform architecture and a famous silhouette clearly visible from the sea.[10]

The later use of the term Skeppsbro nobility to denote the members of the group suggests on one hand their residence, and on the other hand how they defined themselves as a separate social group than the traditional nobility. While they sometimes themselves were ennobled, or married into the old nobility, the new merchant aristocracy more often competed with the old nobility by buying up their landed estates or sometimes taking over their indebted ironworks.[11]

Müller's salon was part of a fashionable French-inspired residence, probably designed by Swedish architects based on Parisian models. The guild painters who made decorations in such homes during the eighteenth century primarily appears to have worked for the wealthiest merchants. The houses inhabited by this newly rich group were divided into different floors, where the salon formed the main room on the second floor with rooms primarily for socialization and entertainment.[12]

It was not the luxury itself or the level of splendour that made Müller's residence or his other properties stand out from the properties owned by his future aristocratic son-in-law.[13] What set them apart instead was the different range of luxury goods that they contained. Müller's clothes and furnishings testify to new habits and customs related to the global colonial merchandise trade as well as products produced in the new manufactures. This included textiles, sugar, tobacco, furniture, playing cards and all sorts of minor objects, which was marketed to the new consumers through advertisements in local newspapers.[14]

Much of the merchandise in the new homes had been produced in Stockholm. By the mid-1700s numerous Stockholm silk weaving manufacturers produced fabrics from imported silk based on patterns and fashion with inspiration from the foremost international models in Lyon and China.[15] The shops in Stockholm, modelled on international forerunners, offered local goods as well as consumer goods from all over the then known world in the same way as in other cities in Europe with access to global trade networks. A Swedish business lexicon that was published in 1815 was so extensive that it had to be printed in two parts to accommodate the international product range and the local variations in colour and design.[16]

The new luxury goods were not only purchased by wealthy merchants. The new luxury consumption was primarily driven by a growing urban middle class: craftsmen, manufacturers, civil servants, minor burghers and retailers.[17] Their homes and home furnishings, however, did not follow the French-inspired cultural pattern that was taken up by the wealthiest group of merchants like Müller, but were of simpler designs.[18] The new type of luxury consumption was still everywhere. Even the poor bought factory-made linen or sugar that had been produced in the slave plantations in the Caribbean.[19]

The members of the Skeppsbro nobility, who were the primary drivers in the process, were also the main exponents of the new luxury. When compared to the old nobility, which inherited its valuables – jewellery, golden leather wallpapers, weapons, tapestries and furniture – the luxury consumers in the new merchant aristocracy represented something fundamentally new.[20]

The painting of Jürgen Müller's salon and the probate inventory shows new modern goods and customs and striking new habits. The people in the painting are drinking tea, surrounded by new goods and materials. Tea was an important part of the new luxury and spread to the well-to-do in the capitals of Europe from the latter part of the seventeenth century.[21] The ladies are reading the newspaper while Müller is talking to one of the foremost officials and lawyers of the time, still wearing comfortable night clothes. The writing paraphernalia is not a historical artefact, but illustrates the ongoing activities of the merchant house with correspondence and accounting. The furniture is of mahogany, an English style-influence that was established around this time.[22] Behind the ladies, a bookkeeper stands ready for instructions that will follow from the conversation between his employer and the initiated civil servant.[23] Jeanette, the future bride, stands in front of the group. In the painting her presence and the dowry she carries with her represent the transfer of the accumulated property to a new generation through marriage.[24]

French influences

The French influence on Swedish culture in the eighteenth century has been thoroughly investigated by researchers with a focus on architecture, artisanal production as well as literature and art more generally.

Sumptuary laws have also been highlighted with reference to Northern Europe in more recent works.[25] In the following I will instead highlight the French influence on three areas that have received comparatively less attention, namely fashion-producing manufactures, fashion magazines and advertising.

Manufactures and fashion

The rise of manufactures in Sweden was intimately linked to the institutional growth of the early modern state during the sixteenth and seventeenth centuries. Today, two main ideas can be distinguished within institutional theory formation; two diametrically different interpretations of the role of pre-industrial institutions in the economic and social transformation of early modern European society.[26]

The first position is based on a positive view of the dismantling of the harmful medieval guild-system, and in its wake, the emergence of a new type of free market, a process which mainly took place in Northwestern Europe, in England and in the United Provinces.[27] The freedom is contrasted with a negative view of the development, in central, southern and especially Eastern Europe where a stifling guild-system reigned preeminent, or further east still where, the so-called "second serfdom" emerged. Several proto-industrial regions were de-industrialized at the end of the early modern period as a direct result of such harmful institutions with a resulting delay in industrialization.[28]

The second main idea on the contrary highlights the importance of guilds and hallmark courts and emphasizes the importance of early modern institutions, stable rules of play and secure ownership during the early commercialization. The framework suggests that the various early modern institutions, interest organizations, welfare systems and complicated ownership rights have been grossly underestimated in research and that these organizations and institutions on the contrary in various ways contributed positively to long-term economic and social development.[29]

Manufactures was one of the institutions, which was intimately connected to the rise as well as the demands of the centralized state. This can be seen both in Philip II's Spain and later in Louis XIV's and Colbert's France in the latter part of the seventeenth century as well as in the expanding Nordic states Denmark and Sweden.[30] In many other continental principalities special privileges were also created for manufactures, especially among the states in the German language area and in the Habsburg Empire.

At the same time, we have to acknowledge that there were big differences between the countries. The demographically sparsely populated and peripheral Nordic countries, where policies on manufactures only were fully developed at a comparatively late point, were in many ways different than Europe's leading mercantile major powers both in terms of demographics, urbanization and general organization. This also means that manufactures, irrespective of if they are viewed in negative or positive terms according to the two above-mentioned main ideas about early modern institutions, must have had different kinds of importance for the emergence and spread of fashion simple depending on geography and time.

The emergence and implementation of state policies on manufacturing in Sweden formed part of a new commercial focus. This process was initiated in the early seventeenth century but culminated in the eighteenth century with a particularly active state economic policy after the end of the Great Northern War in 1719. The rise of manufactures in Sweden happened at the same time as burgeoning world capitalism and an increasingly globalized world trade with which it was closely linked. The manufactures played important roles as fashion distributors in Sweden already in the seventeenth century. The new type of production helped to transfer an international product range of fashion goods and an associated world of ideas into the undeveloped economy. This followed in line with the development in France. In the Netherlands and England the implementation of fashion was probably influenced by other factors.[31]

A good example of how the development affected fashion can be found in the utilization and promotion of colours. In Sweden, state policies on colouring in the eighteenth century on one hand, aimed to build up a domestic production of dyes, and on the other hand, to streamline and organize the production in an efficient way. The education of dyers was supported through state bursaries, and existing international connections were further promoted through travel grants. The results were far reaching, with colours affecting not only the dyeing of fabrics and garments but also the colouring of furniture, glass, porcelain, faience and wallpaper.[32]

Previous research has noted the institutional dimension of the state policy, and in particular, the way in which it challenged the existing guild-system. After the introduction of a new manufacture policy in 1739, guild-appointed dyers were mandated to work under the auspices of so-called "hallmark courts" – a new overseeing administrative institution. Cloth shearers, who processed the felted woollen fabric, were likewise forced to abandon their guild. This probably also applied to the artisans responsible for the decoration of silk and yarn fabrics, a specialist skill where colouring was combined with various textile processing techniques. The process was slow, however, and the new system was met with great resistance.

The new state policy was at times very detailed, almost a little French in its elaborate regulation of the processing process.[33] Instructions detailing how the work should be undertaken in every step were often elaborate. The preparation of cloth was affected by the basic technical difference between short and long fibres. Carded wool fibres and short cotton fibres would be crushed and felted, whereas the parallel long fibres in combed wool and camel yarns were plaited. Of silk, a yarn was made from the fine threads harvested from the silk cocoons. The preparation of short-fibre fabrics began with milling where the fabric was processed by hammering, which resulted in a substantial strengthening of the material. The residual loose fibres were then roughened with special carding combs and the surface lint was removed with specially designed scissors. In the wool industry the latter step was called shearing and much like colouring was considered an advanced craft and a specialist profession. The finer the fashion quality, the more rounds of such finishing were required. An important part of the imitation of international models and qualities was to ensure that the fabric was given the required shine and lustre. While fabrics based on short fibres could be dyed at various stages of the processing, both as unrefined wool, spun yarn and as a finished fabric, the dyeing of silk and combed wool fabrics had to happen during the preparation stages. In connection with this, the fabric would be washed and stretched, and as a final step, pressed.

The many conflicts that arose when errors occurred during the different processing stages dominate the preserved protocols of the hallmark courts and provide a unique insight into the implementation of the colouring policy.[34] Overall, the policy serves as an example of a state-introduced measure that affected fashion, notably by delineating and stipulating the required quality standard through specific regulations.

Fashion magazines

Swedish fashion started to be presented on printed plates and discussed in specialist fashion magazines and printed brochures beginning in the 1810s.[35] The texts became a key place where tastes and perceptions of what was to be perceived as quality were formulated. Some 40 brochures and magazines related to fashion are known from Sweden during the early nineteenth century. About half of the magazines explicitly addressed female consumers with a content that was focused primarily on clothing as well as embroidery.[36] Almost three quarters of the total number were issued in Stockholm. Most magazines only circulated for a couple of years before they were closed down, and only a quarter survived for more than five years. Overall, the publishing of fashion magazines declined in the 1830s, at about the same time as advertising in the daily press became more advanced and differentiated.

Most of the fashion magazines in Stockholm were founded by booksellers. Booksellers in general were both important initiators of advanced marketing techniques and regular advertisers. This meant that they could actively influence fashion by promoting ideas of style and taste and to this end designed magazines that targeted specific consumers. In this process, they also transmitted foreign impulses. Several fashion magazines were in fact straightforward translations of foreign journals.

In the following I will present an analysis of the fashion magazines that were published during the early nineteenth century primarily based on three original publications issued in Stockholm: *Sofrosyne* (1810s), *Magasin för Konst, Nyheter och Moder* (1820s) and *Stockholms Mode-journal* (1840s).

The fashion magazines in Sweden during this period generally conveyed a view of garments focused on colour, appearance and uses. Other aspects of fashion, such as clothing accessories and decorations, only appeared in isolated cases. The older magazines were often dominated by other contents, with fashion presented towards the end under a separate heading. Towards the end of the nineteenth century, the magazines became more varied and consistently included more coloured plates, showing luxury garments for special occasions or, in general, for the well-heeled.

Parisian fashion was the fully dominant reference. England and Germany were only mentioned infrequently. Clothes for women dominate and were also generally more varied than the male garments that were presented. Garments for children were sometimes included, but overall remain rare.

Wool was the prime material and woollen cloth the main fabric in descriptions of tailoring in the magazines. Apart from outer coats and mourning dresses for women, woollen cloth was almost exclusively a fabric used for men's clothing at the time. This included tailcoats, *bonjour* coats, surtout, jackets and various other French styles, trousers as well as hats. Descriptions of woollen cloth for trousers died away after 1825 when camlet (a worsted fabric) took over.

There is continuity in eighteenth-century fashion with regard to French designations of the most important coats and trousers in men's clothing. Tailoring information from the second half of the century therefore provides some guidance for how the design, material, sewing and dyeing of the most important fashion elements was conducted in the early nineteenth century.

The most exclusive and expensive type of garment was the tailcoat.[37] The term both referred to the entire attire (coat, vest and trousers) but also simply to the coat. The coat and trousers by the end of the century were typically black or very dark, midnight blue.[38] The coat, which was life cut with a front and back skirt, was according to the older fashion magazines made from woollen cloth. The lapels were pointed with a silk front. The back was sewn with hook seams in the back between the sleeves.[39] Since the coat was worn open, a so-called *soutach* ribbon was used to hold the front pieces together.[40] The tailcoat trousers were made from woollen cloth or cashmere, a high life cut, silk stripes that were sewn along the outseam of the trousers and worn with braces.[41] Cashmere, which is often mentioned in the older fashion magazines, was a softer woollen fabric than normal woollen cloth that was more finely woven with three shafted weft twill. The yarn was spun from the fine wool taken from Indian cashmere goats.[42] The tailcoat was usually worn with a white vest without back, made from *piqué*, a double-woven white cotton fabric with a pattern of small squares. The same fabric was also used for the collar, the shirtfront and the tailcoat fly.[43] The shirt would be white with a reinforced *piqué* shirtfront (with neckband and a neck warmer attached with shirt buttons), reinforced cuffs and a collar with folded ears. Shoes made from patent leather or *chevreau* shoes finished off the ensemble.

The *bonjour* coat was a life-cut frock coat (as was the cheaper habit coat), according to the fashion magazines of the 1820s usually made from black fine woollen cloth with straight knee length skirts. It was almost as expensive as a tailcoat. A distinction was made between one-row and two-row, as well as trimmed and untrimmed edges, just like for the tailcoat. Two-row meant that the garment had buttons and buttonholes on both front pieces, which appeared as two lines when the *bonjour* was closed. It was sometimes also called *redingot* or long coat. It could also be made from worsted wool (a slightly thicker quality of double-twisted yarn) and then became cheaper to manufacture. The back and front skirts were often sewn with folded flaps with silk lapels attached to the front piece to form the chest portion.[44]

Fashion changes in men's clothing can be divided into three levels: short, medium and long-term changes. Short-term changes happened as part of the regular tailoring and fitting of a new garment, a process during which the piece was adjusted to fit the body of the customer. This could still be influenced by fashion shifts. According to an article in *Sofrosyne*, tailors complained when they sometimes had to cut and recut the same garment six times, to a point where they in the end were back to their starting position. Medium-term fashion changes refer to adjustments made to existing garments, for example, by adding new collars, buttons, changing the liner or adding new vests. Long-term fashion changes finally meant the purchase of newly designed garments of the latest model and colour.

The use of colour seems to have become more advanced in the 1820s. In *Sofrosyne* black or blue cloth was the typical colour for coats. During the 1820s, the variation became larger with blue, white, brown, black, green and mixed shade colours appearing in the descriptions and plates in *Magasin för Konst, Nyheter och Moder*. The *Stockholms Mode-journal* paints a similar picture of the situation in the 1840s. By the end of the period of investigation, plates in the fashion magazines, now increasingly in colour were getting more common, and the texts more detailed.[45]

An interesting shift happened in the way in which women and men's clothes were produced, during the period. In the eighteenth century female dresses were still made by highly skilled guild-connected male tailors. Pernilla Rasmussen has shown that the female dress changed in a fundamental way between 1770 and 1830, however, as a result of a growing transformation of the gender division of labour. The process saw a gradual division, where male tailors handled male clothing, while a smaller group of female tailors started making clothes for both genders. This led to a further change, which started around the turn of the 1800s, which witnessed female seamstresses expanding traditional home sewing to gradually taking over the general sewing of women's clothes, a transition which was completed by the 1830s.[46] The change was multifaceted and should not be further developed here. It could be noted how the development was similar to the situation in France. In a context where a growing number of unmarried women entered the labour market, an incentive to accept that women's tailoring should be handled by women was provided. In the context of the present investigation, the development is obviously interesting, when considering the way in which female dress was promoted and illustrated in the early nineteenth-century fashion magazines. The plates in the new magazines thus not only, as initially mentioned appealed to female consumers, and spread French influences but in this process also essentially supported female dressmaking.

By the mid-1800s the communication with a new urban female readership was becoming more apparent. A plate from the 1844 issue of *Magasin för Konst, Nyheter och Moder* shows three ladies around a red sofa. Two of them look at the reader while the third one stands behind them, leaning lightly against the couch edge and showing off her richly decorated hat. When considering how the material for the dresses, ribbons and hats that were displayed in plates in the fashion magazines also was made, among other things, at silk manufactures in Stockholm, where most of the qualities were based on French patterns, it becomes clear how the magazines helped in promoting the use of French fashion and imagery.

Advertisements

The early French-inspired fashion magazines, with relatively limited print runs, probably only reached the urban elites. It was only through advertising in the daily press that the luxury that was presented in the fashion magazines was developed into a new luxury and diffused to a wider market and a broader middle class readership. The process happened after traders and manufacturers started transferring, incorporating and transmitting the quality designations from the fashion magazines into their advertisements for textiles and subsequently clothing as well as accessories.[47]

Research on Western European and American conditions indicates that most of the early advertising was done by merchants. An early ability to capture new marketing opportunities has been seen above all by manufacturers and textile traders. In both Great Britain and Germany, manufacturers and textile traders were the pioneers when older forms of distribution developed into more specialized trade such as fashion stores, bazaars and department stores with a higher degree of fixed prices and standardized sales. Already at an early stage, advertising in the daily press seems to have been of great importance.[48]

Advertising and articles on clothing and fashion emerged as early as the late eighteenth and early nineteenth centuries. The emergence of periodic publications (daily press, newspapers, magazines) was an important part of this development. In Great Britain, during this period, special fashion magazines for men and women became more numerous. The information was disseminated into the countryside, mainly with the help of travelling traders. Rural newspapers emerged in the eighteenth century. In the latter half of the century, advertisements seem to have become an accepted method for local marketing as well. In the beginning, it was primarily booksellers that accounted for the advertising, which likely was related to the fact that bookstores often printed the newspapers. Books, quacksalver drugs and real estate dominated the early advertising completely.[49]

In Sweden, during the first half of the nineteenth century, newspapers were also increasingly founded outside the larger cities of Stockholm, Gothenburg and Norrköping and the university towns of Lund and Uppsala. In Stockholm, two substantial newspapers were published six days a week already in the eighteenth century. Otherwise, at this time, newspapers were often simpler four-page magazines that were published between once and twice a week. These features remained in principle during the first decades of the nineteenth century. In the 1830s, an uninterrupted period of growth began, which meant that more magazines were added in traditional publishing places and that magazines began to be published in new geographical areas. The rural newspapers before 1850 usually appeared once or twice a week and consisted of four pages with two to three columns. Advertisements for goods typically occupied one half to one column in each issue. For the newspapers that had been established before the 1840s, there was usually a growth in the total amount of advertisements in the 1830s and 1840s.[50]

Prior to the 1830s, advertisements for colonial goods and factory-made consumer goods dominated in important newspapers in Norrköping and Stockholm. Substantial textile advertising, meaning adverts on textiles or similar wares beyond single garments, became more common in the 1830s and 1840s. The advertisers were comparatively advanced and dynamic in their approach, linking fashion, textile qualities, assortments and price. Only a minority of the adverts for textile products had no connection at all to fashion.

Almost half of the advertisements that have been investigated promoted the products as "modern" or "of the latest Paris fashion" or similar statements of origin, novelty and quality. A further group included advertisements that linked textiles to fashion by using terminology on textile qualities identical to those reported in fashion magazines or in textile trade publications. Together, these two groups constitute an overwhelming majority of the total number of textile advertisements.[51]

Conclusions

This chapter has shown how French fashion ideals and ideas had a big and broad impact in Swedish early nineteenth-century society. On the basis of Maxine Berg's conceptual dichotomy between old and new luxury I have argued how this development followed a transition from early modern to modern luxury consumption. During the period state-supported manufacturers increased the production of textiles and clothes; fashion magazines produced a fashion discourse; and traders started advertising in the daily press. French models for clothing and accessories as well as quality designations in French grew increasingly important as a result and helped to recreate and diffuse a French fashion image in Sweden.

The results thus suggest how this image was the result of a convergence of both public and private institutions and private entrepreneurs on a growing increasingly modern market that emerged in the first half of the nineteenth century. On this market state-regulated manufacturers used French ideals and textile qualities when producing silk and woollen textiles and the emerging fashion magazines presented exactly how these textiles could be used fashionably. This information, in turn, was taken up by merchants who introduced the information and placed it in a wider context with further types of goods, when advertising in the emerging daily press. The result was the spread of a new type of luxury in Swedish society.

The development should be seen against a general, much more well-known shift during the eighteenth century which saw French aesthetics and culture influence Swedish architecture, interior design, art and language in a very pervasive way. This old luxury was based on long-term credit arrangements. In the new development that emerged from the second half of the eighteenth century, this was supplemented by a growing market for promissory notes, which enabled the consumption of new luxury and the type of new fashion that emerged during the period.

The development can also be seen in relation to the ideas about an industrious revolution as presented by Jan de Vries. Even though this theory is based on a much earlier development, it is still reasonable to assume that fashion magazines and advertisements in the daily press stimulated a growing desire which, in turn, affected the way in which people viewed their household needs and developed their consumption patterns.

Ideas of novelty, and quality repeated in various media and visualized on plates in fashion magazines, combined with shifts in fashion changes, most likely affected people and created a new desire for taste and a new lifestyle that could be manifested in design, decoration and garments.

Notes

1 Riello and Rublack (2019), for sumptuary laws. Stavenow-Hidemark and Nyberg (2015) for the Swedish eighteenth-century textile manufactures. See also Henderson (1985); Braudel (1986); Molà (2000).
2 See note 36 in the General introduction.
3 Cole (1939); Heckscher (1994[1935]); Nyström (1955); Braudel (1986); Ogilvie (1997).
4 Kjellberg (1943); Nyström (1955); Westerlund (1988).
5 SSA, Storkyrkoförsamlingens kyrkoarkiv, EIa, Lysings-och vigselböcker, huvudserien, vol. 6, March 30, 1811. The following is partly based on my contribution in von Wachenfeldt and Nyberg (2015), pp. 9–13, 16.
6 Forsstrand (1918), pp. 160–177. The painting is also reproduced in Cederblom (1929).
7 Forsstrand (1918), p. 160.
8 Forsstrand (1918), p. 172f.
9 de Vries (1994, 2008); Söderberg (2007), pp. 329–332.
10 For information about the history and role of the Skeppsbro area, see Nyberg (2006), pp. 22–38.
11 For old and new luxury, see Berg (2005), ch. 1; For the Skeppsbro nobility, see Nyberg (2000).

12 Danielsson (1997), p. 118f.
13 Forsstrand (1918), p. 165.
14 Nyberg (2010), p. 95f.
15 Nyberg (2013), p. 121f.
16 Synnerberg (1815), parts I–II.
17 Nyberg (2010), p. 95f.
18 Danielsson (1997), p. 118f.
19 Söderberg (2007); Müller (2018).
20 Berg (2005), ch. 1.
21 Ahlberger (1996); Styles (2007).
22 Sylvén (1996), p. 26. See also Snodin and Styles (2004a, 2004b), pp. 151–153; Hodacs (2016).
23 Forsstrand (1918), p. 166.
24 Erickson (2005[1993]), ch. 5.
25 Riello and Rublack (2019), part I.
26 The following section is partly based on Nyberg (2016), pp. 338–350.
27 This type of "free" market is related to the emergence of an industrious revolution in these areas as early as the late seventeenth century according to Jan de Vries. This approach, as mentioned, theorizes the changed resource allocation of households as a result of a growing demand for consumer goods. Cf. with McKendrick et al. (1983); Berg (2005), ch. 1.
28 Ogilvie (1997, 2003, 2019).
29 Epstein (1991); Greif (2006).
30 Braudel (1997); Cole (1943).
31 Kjellberg (1943); Nyström (1955).
32 Kjellberg (1943), ch. XI.
33 Nyström (1955), pp. 57–59.
34 Nyberg (1999b), ch. 3.
35 The following section is based on Nyberg (1999b), pp. 130–132, 218–220.
36 See Nyberg (1999b), pp. 218–220.
37 The tailoring cost in 1882 in Stockholm was between 19 and 21 Swedish kronor, depending on the level of tailoring, for an estimated 60 hours of work. In Gothenburg the cost was between 16 and 18 kronor. Björkman (2004), p. 100; Hjern (1943), p. 43; Komiterade för Stockholms skrädderiarbetareförening april 1882.
38 Björkman (2004), pp. 28–30.
39 Björkman (2004), p. 108.
40 Björkman (2004), p. 97.
41 Roetsel (1999), pp. 326–327.
42 Preliminär textilteknisk ordlista (1957), p. 116.
43 For the technique to produce the pattern, see Preliminär textilteknisk ordlista (1957), p. 172.
44 Björkman (2004), pp. 19, 56, 82.
45 Nyberg (2015); Cizuk (2015).
46 Rasmussen (2010).
47 Nyberg (1999a).
48 Benson and Shaw (1992), pp. 31, 137, 139, 161; Fine and Leopold (1993), pp. 5, 83, 91f.
49 Chandler (1977), pp. 167–174. On advertising, see Lemire (1991), p. 163f. For fashion magazines, cf. with McKendrick (1982), p. 41f.
50 Nyberg (1999a).
51 Nyberg (1999b), ch. 5.

Part II

The Swedish financial system and bankruptcy law

3   The Swedish bankruptcy system, 1734–1849

Karl Gratzer, Mats Hayen and Klas Nyberg

Introduction

In the mid-nineteenth century, the Swedish credit market was still dominated by private financial networks and by a private supply of capital. Banks, discounts and other public financial institutions only played minor roles as financers of trade and industry. In spite of these circumstances private financial networks were embedded in public financial institutions and bankruptcy legislation. In Sweden during this period, the institutional administration of bankruptcies was gradually developed to further minimize the damage from personal insolvency. This was not a unique development, but on the contrary happened across the whole of early modern Europe as well as in colonial America.[1]

The reformation of the Swedish bankruptcy institution during the eighteenth century led to the bankruptcy process becoming more efficient. The underlying aim was partly to realize the estate's remaining assets and divide them between the claimants and partly to ensure that none of the creditors were allowed to profit at the expense of the others through compulsion and manipulation. The legal reforms helped to diminish risks and losses for merchants and private bankers during bankruptcies. The reforms also played a crucial role in the development of the financial market and helped to increase the supply of capital during the period of early industrialization in the beginning of the nineteenth century.[2]

In this chapter we will describe the development of the bankruptcy system in Sweden between 1734 and 1849, placed in a European perspective. When considering how the aim of the book is to discuss and problematize the transformation of early modern ideas of credit and fashion among luxury producers and how credit and trust among those groups were affected by shifting economic behaviour and situations, it naturally also becomes crucial to understand the reformation of the bankruptcy institution.

The historical background

The origin of the concept of bankruptcy lies in the concept of credit. When an individual applies for credit or borrows money, he or she enters into some form of written or oral agreement. If the repayment is not made, the debtor violates a fundamental contract, a violation of ownership. For many centuries, a debtor who was unable to repay his debt was treated with harshness and the inability to pay was often equated with theft.

According to old Roman law, as inscribed in the Law of Twelve Tables, anyone who wanted to take credit offered himself, his family and his property as collateral to the lender. If the person who had borrowed failed to fulfil his payment obligations, he became the property of his creditor, who even was allowed to kill him. If there were several creditors, the law stipulated that they were entitled to cut the debtor's body into equal pieces.[3] In this cruel custom, we can trace the first hint at a bankruptcy legislation. Indeed, the purpose of the bankruptcy institution is to achieve justice by preventing competition between creditors. It is unclear to what extent debtors were actually dismembered. The debtor's inability to pay always led to bondage, however, and to the possibility that his family would be sold as slaves. The idea of bondage to repay debts remained a viable concept well into the medieval period.

The Roman bankruptcy legislation ceased to function in conjunction with the disintegration of the Empire and its institutions. For a long time thereafter, a long-standing bankruptcy system was missing in Europe. The Germanic tribes brought their own right-system into the Roman territories of Gaul, Italy and Spain, where they often lived alongside the older, Romanized population and its laws. Germanic law has been described as a customary law, a product of the customs of the people, which had been preserved as an oral tradition, in the memory of juristic men. Nordic law was essentially a branch of Germanic law which, among other things, was expressed in Swedish provincial laws in the Middle Ages.[4]

In Germanic law, the insolvent debtor was subjected to a similar harsh treatment as in ancient Rome. One of central ideas was that the inability to pay a debt equated to theft. If the debtor lacked assets, he was much like in Roman law given to the creditor as a slave. Prison, even torture, was used as a means to obtain property.[5] Unlike in Roman law, there was no distinction between honest and fraudulent debtors in earlier Germanic law. The foreclosure always initially aimed at the debtor's fortune, but unless this was proposed, he was handed over to the creditor as a slave and could be sold or killed.[6]

It was not until the Late Middle Ages that a new legislation on bankruptcy was developed by emerging royal powers and the church.[7] The main source was *Corpus juris civilis*, a Roman collection on civil law, dating from the sixth-century AD.[8] The Roman bankruptcy law generally saw a revival in Middle Age Europe, where in particular, Italian trading cities and the associated credit markets that emerged during the thirteenth century demanded new commercial legislation.[9] Strong impulses from the older legislation spread across Europe, which greatly affected the perception of how indebted people should be treated.[10] Only debtors who could prove that they had ended up insolvent due to an accident would escape punishment. The withholding of property or attempts to escape from creditors would be punished with anything from losses of rights, jail or even a death sentence. In many cities, voluntary surrendering of property, so-called *cessio bonorum*, was completely excluded.[11] The harsh attitude, essentially based on the assumption that every insolvent debtor was deceitful and should be treated accordingly, spread across Europe, especially to France, Spain and England.[12]

The essence in medieval Italian bankruptcy law was well captured by the Italian jurist Baldus de Ubaldis in the declaration of *fallitus ergo fraudator* (insolvent and thus an impostor). Experience had often provided support for such harsh judgements, and "fraud" was, as a result, suspected in almost all bankruptcy proceedings. This led to another postulation *falliti sunt infames* (insolvency brings dishonour). These two judgements cast their shadow over the bankruptcy legislation in Europe well into the nineteenth century and form the basis for explanations of what caused insolvency.[13]

The growth of the bankruptcy institution from the late Middle Ages onwards was closely linked to the development of commercial capitalism.[14] Bankruptcy legislation was developed as part of the commercial legislation in Italian city-states, but subsequently spread to the new economic centres that developed in Northern Europe, including London, Hamburg and Amsterdam.[15]

During the late sixteenth and early seventeenth century, Antwerp led the way in the formulation of commercial legislation in the Low Countries.[16] As part of this, the Flemish town also became an important forerunner in a development that saw bankruptcies shift from being viewed as a wrongdoing that was covered by criminal law to being regarded as an issue that was handled under commercial law.[17] Following the independence of the United Provinces, the economic power centre in the area shifted northwards. In Amsterdam, a special commercial court for insolvency cases, the so-called *Desolate Boedels Kamer* was established in 1643, which would form the model for a number of similar courts in other Dutch cities.[18]

Overall, in countries with strong state power, such as the Spanish early modern state, commercial law fell under public law. This approach, in which the entire insolvency problem was treated as a legally coherent whole, inspired and subsequently spread further to German-speaking principalities.[19] Countries with weak state power and fragmented legal systems in contrast were characterized by a general, vaguely designed bankruptcy law consisting of a slowly growing collection of overlapping regulations and local features. Here, creditors had a comparatively greater influence, and bankruptcies fell under private law.[20]

From its origin as a narrowly delimited special legislation for merchants and merchant activities, commercial law was gradually expanded to cover broader sectors of the economy.[21] The development happened in many of the early modern state across Europe with a particular active period in the late seventeenth century. From having been fairly uniform and similar in character, differences now also arose in the orientation and formulation of bankruptcy legislation. In France, for example, a more detailed regulation followed when commercial law was codified under the *Ordonnance du Commerce* in 1673.[22] The regulation made the legislation more coherent. The combination of Italian commercial law and royal decrees still meant that the bankruptcy law created legal controversies when the new laws collided with existing legislation.[23]

The differences in the bankruptcy legislations were the result of different general economic policies where the interaction between the state and municipal authorities and institutions was of particular importance. This more specifically included input from various legal and administrative institutions, including general commercial courts, insolvency courts as well as judges and lawyers at state courts. These issues were in turn linked to the constitutional histories and outlooks of different countries. Financial theories and other arguments led different countries in different direction, with German-speaking areas for example being heavily influenced by cameralism while English reformers were inspired by Daniel Defoe and similar writers.[24]

The Swedish bankruptcy legislation 1734–1818

At the end of the seventeenth century a reformation of the existing Swedish legal system was initiated. The process reached an end point in 1734, when the first comprehensive Swedish Civil Code, divided into nine separate laws, was published. The new code replaced all existing provincial laws and city laws, thus creating a unified countrywide system.

The rules on bankruptcy in the new Civil Code were brief, but nonetheless replaced a number of even briefer medieval decrees and bankruptcy regulations. Chapter VIII of the Debt Enforcement Law dealt with "sequestration and debtors' prison", detailing how the debtor's property could be held as security, so that it could be used to pay his debts, and how the debtor could be deprived of his freedom because of the debt. Chapter XVI of the Commercial Law stipulated that the default debtor should be imprisoned if it was found that his "poverty was due to wastefulness, gambling, idleness or carelessness".[25]

The influence from Roman law was still very much evident. This for example was the case with *cessio bonorum*, i.e. the debtor's protection from creditors after having voluntarily surrendered assets. The chance to make a new start after this type of ruling had been present in Swedish law at least from the early seventeenth century. The 1734 Civil Code instructed that if the debt was a result of "sea damage, military enemy attack, fire or other accidents", the court could rule that the debtor, after satisfying his creditors with his current assets, should be freed from all future claims. Similar statements were subsequently included in separate bankruptcy laws published in 1773, 1798 and 1818. Each new law, however, made it harder to meet the requirements connected to the right, and the ability to refer to it was finally completely removed in the bankruptcy law of 1862.[26]

Debtors' prison, which had previously been a safety measure to prevent debtors from escaping from their liabilities, in 1734, became an instrument that could be used on people with debts connected to unsettled bills or overdue promissory notes. Debtors' prison in short became an alternative for creditors as soon as a debtor failed in his obligation to pay. If a debtor after this filed for bankruptcy and the court accepted the application, the threat of debtors' prison still remained. This fact would not change until 1766, when a new important ordinance regarding bankruptcy came into effect.[27]

The regulations regarding bankruptcy remained in force until 1750, after which several new ordinances were introduced. These primarily tried to come to terms with the manipulation of the use of the debtors' prison, and more specifically, the way in which imprisonment could become beneficial for the debtor. In times of war, soldiers could for instance escape military service by serving in prison for debt.[28]

The next important change to the Swedish bankruptcy regulation came after a major trade crisis, which started in Amsterdam in 1763, and quickly spread across Europe. The Swedish financial market had strong ties to Hamburg, London, Amsterdam and Berlin – the most important economic centres in Europe – and was soon directly affected.[29] The turmoil, and a large number of bankruptcies, resulted in the state taking action to improve the legislation to better handle the situation. This resulted in the publication of two new bankruptcy ordinances in 1766 and 1767.[30] This was followed in 1773 by a bankruptcy law, the first to declare that the efficient handling of bankruptcies was an important part of commercial life in a nation.[31]

The new legislation changed the way in which the debtor was treated during the bankruptcy proceedings. Prior to 1766 a bankruptcy was in most instances devastating for the debtor. The principal aim of the bankruptcy was to extract as much as possible from the bankrupt estate and to distribute it among the creditors, leaving the debtor without much to benefit from the proceedings.

This changed with the new bankruptcy ordinances of 1766 and 1767. Together, they introduced two new concepts: *protection* of the debtor during the court proceedings and *preclusion* on the day of proclamation, two important steps towards a modern way of handling insolvencies and bankruptcies in Sweden.

The idea of protection was introduced in the 1766 ordinance. Specifically it protected the debtor from distraint, or in other words from the sequestration of personal property during an ongoing bankruptcy from the day of the bankruptcy application until the verdict. This had an important impact not only for debtors in future cases but also for the administration and organization of older cases. With the new understanding that the debtors would be protected until a verdict was reached, it suddenly became very important for the court system to bring still unresolved cases to trial. In Stockholm this resulted in the compilation of a register of all unresolved bankruptcy cases, which in the end covered a remarkable 900 cases stretching back as far as 1707.[32]

The idea of preclusion on the day of proclamation was introduced in the 1767 ordinance. It stated that the debtor and all creditors in a case of bankruptcy had to present their claims before noon on the day of proclamation or in other words on the day when the official bankruptcy proceedings got underway. Failure to do so resulted in a loss of rights to speak and present claims during the subsequent negotiations. Preclusion usually had a positive effect for the debtor, as all debts to creditors who failed to present their claims in time were renounced. After 1767, debtors in cases of bankruptcy could often see between 10 and 50 per cent of their debts disappear simply because of the disqualification of their creditors.[33]

The combined changes that were implemented between 1766 and 1767 had important effects on the actions taken by both debtors and creditors. The protection from debtors' prison, from possible seizure of property by individual creditors, and the preclusion on the day of proclamation were beneficial for the debtor. Applying for bankruptcy as a result became a rational choice for many debtors after 1767, as it for the first time could lead to positive results.

Swedish laws regarding bankruptcy at this time mainly followed the German tradition, which in many ways was different from contemporary French and English legislation. An important principle, with its roots in the German tradition, was the unity of the bankrupt estate. This principle became central in the 1773 bankruptcy law. The bankrupt estate, which was clearly defined by the court at the start of proceedings, was administered by two elected curators, selected from among the group of creditors. The bankruptcy process followed clearly defined steps from the day of application until the verdict, which included setting down the order in which the assets would be divided among the creditors.

The control of the duration of the bankruptcy cases became important in the new system, as the debtor, as mentioned, in many cases would benefit from long proceedings because of the protection against distraint and sequestration that existed until the court reached a verdict. Protracted cases also meant that creditors had to wait for a long time before getting their money back. An average bankruptcy case in the 1780s took approximately 1,000 days from the day of application until a verdict was delivered. Several measures regarding bankruptcy procedural in the late eighteenth century reduced this time to around 500 days in the early nineteenth century (see Figure 3.1).

The need to reduce the time was enforced in an ordinance in 1768 and was also emphasized in the bankruptcy laws of 1773, 1798 and 1818. The law of 1818 clearly states that in cases where "[...] the right of the majority of creditors, will become delayed, due to complaints regarding a single claim, with hearings of suitors of witnesses [...]", the court could separate contested parts of the case and proceed to rule on the rest.[34]

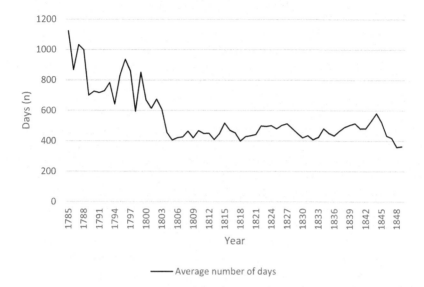

Figure 3.1 Average number of days from application to verdict for bankruptcy cases begun in the given year, 1785–1849

Source: SSA, Stockholms magistrat och rådhusrätt, C5a, Konkursdiarier, 1785–1849.

The continued modernization of the Swedish bankruptcy law 1818–1830

The slow handling of bankruptcy cases continued to trouble the authorities as the nineteenth century progressed. Many debaters argued that the traditional way of handling bankruptcy cases as ordinary civil cases in a German tradition was outdated and that the French system, regulated in Napoleon's *Code de Commerce* from 1807, was better suited for a more efficient handling of bankruptcies.[35]

By the late 1810s criticism was also directed at the existing bankruptcy law, in its second, 1798 incarnation, with reference to it being unclear, powerless and generally too accommodating to the debtors.[36] Proposed changes were made, and a new bankruptcy law, the third since 1773, was adopted in 1818.

The new law differed from its predecessors through a more systematic arrangement, more sharply defined provisions, new statutes concerning forced bankruptcy and through a more stringent liability for careless or fraudulent debtors. The law also contained other news, with regard to how the bankruptcy would arise and how the estate would be managed. The new law provided three examples of when a debtor could go bankrupt if he had not voluntarily resigned the property to the creditors. One of the grounds was that the creditors feared that they would not receive full compensation for their claims after the debtor's property had been divested. The second reason was that the debtor's property already was under foreclosure, whereas the third reason was if the debtor had been in debtor's prison for a period of two months or longer.[37]

Provisions were also introduced which defined how property that had been surrendered by the debtor should be managed "with diligence, skill and fairness". The creditor's interest would be safeguarded in a more efficient way, which would be achieved through the issue of ordinances for curators and custodians. The custodians could be given guidelines for how to conduct their administration, and the curators were required to report on the administration on a more regular basis. The court could require the individuals to fulfil their duties under threat of separating them from the assignment. Both custodians and the curators were given a new number of assignments that they were obliged to fulfil, whereas the custodians also were granted compensation for their work.[38]

The rights of wives in bankruptcy cases that involved their husbands were also strengthened in the new law. If an application from a wife to have her assets legally separated from the bankrupt estate was accepted by the court, the required part of the estate was now immediately moved out of the reach of the creditors. The wife, who formally was under the guardianship of her husband and thus not in a position to guard her rights, would also receive special assistance from a custodian at the court hearings.

In the law of 1798, a debtor who filed for bankruptcy could gain certain material benefits. In the law of 1818, these benefits were limited to only include a safeguarded personal freedom. The debtor could no longer apply for cession to be released from future payments. This meant that even a non-negligent debtor who had become insolvent would be liable to repay the debt with assets acquired in the future.[39] If the assets after a deceased debtor were considered insufficient to pay his debts, the heirs had the opportunity to apply to the court to forego the inheritance, leave the entire estate to the creditors and be declared free of all future claims.[40]

The 1818 law also distinguished itself from previous bankruptcy proceedings by its sharply limited provisions, among other things, on liabilities for careless and fraudulent debtors. Negligence was, as a concept, invoked already in the 1734 Civil Code, where it was contrasted with fraud and unintentional bankruptcy.[41] The subsequent bankruptcy statutes also included certain regulations to sanction such negligence. Actual penalties for careless debtors were first included in the 1818 bankruptcy law, however.[42]

It could be argued that a French influence lay behind the sharpened view of the debtor's behaviour that characterized the new legislation.[43] French law attracted a great deal of attention in Sweden in the early nineteenth century.[44] The strict judgement of the bankruptcy debtors, as expressed in the *Code de Commerce*, which entailed a moral condemnation of the inability to pay, was now allowed to dominate also in Sweden.[45]

Overall, the 1818 law was a key step in the path towards the modernization of the Swedish bankruptcy law that had started in 1734. The development of the Swedish economy, in general, and the expansion of business operations, with more and larger credits, in particular, still meant that the new law would need amendments.

In 1830, a new bankruptcy law was announced. One of the novelties was the recognition that less complex bankruptcies could be handled by only one custodian.[46] While the 1818 law had brought about significant improvements to the earlier legislation, the 1830 bankruptcy for the most part only contained minor additions and generally agreed with its predecessor on most points.[47] The new law would be criticized for not being able to change the "damaging slowness" in the treatment of bankruptcies. The provisions regarding the debtor's liabilities were furthermore considered insufficient to be able to prevent abuse of trust, which was deemed necessary for business operations.

One of the reasons for the relatively short life span of the bankruptcy laws during the first half of the nineteenth century was the rapid economic changes that Sweden went through during this period. When the Civil Code was published in 1734, most of the commercial activities in Swedish society, next to the agricultural sector, were operated as private business enterprises. The majority of the business operators were urban artisans and merchants. When they became insolvent, no distinction was made between their role as private individuals and as company owners. In order to satisfy the creditor's requirements, both the assets in the household and in the business were included in the proceedings and made available for payment of the overdue debts.[48] The bankruptcy as a result did not mean that the debtor became debt free. This only happened when the debtor died and if the heirs relinquished their claims on the inheritance.

The final step in the modernization process of the bankruptcy institution happened with the introduction of the limited company. The joint-stock company was a prerequisite for the development of the financial markets that took place during the twentieth century. As in the rest of Europe, the roots of this corporation form in Sweden can also be said to go back to the trading companies of the seventeenth and eighteenth centuries.

It was only with the publication of the Companies Act of 1848 that this new form of company became legally regulated, however. A new law in 1895 marked the transition to more modern legislation where the state power formally relinquished control of individual corporations.[49] The new form of association enabled larger projects to be financed by spreading the risk to several people and limiting them to the invested capital. When a limited company was liquidated in a bankruptcy, the debts, unlike in the previous private business enterprises, also disappeared.

Conclusions

In this chapter we have shown that the reform of the Swedish bankruptcy legislation was part of a wider historical and European development. The period from the late seventeenth century onwards appears as a particularly important period, with more active and intervening early modern states taking measures to begin reforms in many parts of Europe. At that time, the bankruptcy legislation was also extended to other parts of the economy beyond the narrowly delimited special legislation for merchants. From having been fairly uniform and similar in scope, differences now arose with regard to the limitations and scope of the legislation.

Everything indicates that the Swedish legislation incorporated expanded bankruptcy laws during a modernization wave from 1767 until the beginning of the nineteenth century, as part of the international development. During this process, rights and obligations were clarified, responsibilities were defined and a growing realization that cases had to be decided within a reasonable time was addressed through new laws. The modernized legislation in Sweden emerged at the same time as many of the leading commercial nations in Europe under a significant influence from German legislation.[50] In Sweden one of the most important results of this streamlining was, as we have seen, that the processing times of individual cases gradually decreased from about 1,000 days, on average in the 1780s, to around 500 days by the turn of the nineteenth century.

Notes

1 For the Swedish development, see Olivecrona (1862); Ögren (2010); Sallila (2016). For the European and American development, see Welbourne (1932); Duffy (1980, 1985); Hoppit (1987); Markham (1995); Arrunada (1996); Gratzer and Sjögren (1999); Balleisen (2001); Skeel (2001); Mann (2002); Fridenson (2004); Di Martino (2005); Gratzer and Stiefel (2008); Safley (2013); Cordes and Schulte Beerbühl (2016); Kunstreich (2016); Wilmowsky (2016).
2 Nyberg (2010); Hayen and Nyberg (2017).
3 Alexander (1891); Lantmanson (1866); Erler (1990).
4 Nordisk familjebok (1908, 1933).
5 von Amira (1913).
6 Grimm (1881).
7 Amira (1913).
8 Åbjörnsson (2016); Bergström (1771); Olivecrona (1862); Tuula (2001).
9 Sallila (2016), p. 28.
10 Inger (1997).
11 Rydin (1888).
12 Hunt and Murray (2000).
13 Gratzer (2008).
14 Sallila (2016), pp. 35–87.
15 Carboni and Massimo (2013); Sgard (2013); Wakefield (2013); Schulte Beerbühl (2015), pp. 16–17.
16 Sallila (2016), p. 60; De ruysscher (2016); Gelderblom (2013).
17 Sallila (2016), p. 61.
18 Sallila (2016), pp. 62–63.
19 Sallila (2016).
20 Curtis (1980); Rodger (1985); Duffy (1985); De ruysscher (2016), p. 78; Nyberg and Jakobsson (2016).
21 Sallila (2016), pp. 27–33; Duffy (1985), ch. 1; Mann (2002), ch. 2.

22 Sallila (2016), p. 58.
23 Sallila (2016), p. 59.
24 Schulte Beerbühl (2015), p. 9f.
25 Sveriges rikes lag (1734, 1934), pp. 118–119.
26 Agge (1934), pp. 918–921. For the concept of *cessio bonorum,* see Häberlein (2013), pp. 19–26; Reynard (2001), pp. 355–390; Fischer (2013), pp. 173–184; De ruysscher (2013), pp. 185–199. In English bankruptcy laws, the possibility to be liberated from residual debts was introduced in 1705 in the so-called "Act of Anne". Sgard (2013), pp. 225–235.
27 Gratzer (2008), pp. 15–59; Nyberg and Jakobsson (2013), pp. 72–93.
28 Kongl. Maj:ts nådige Förklaring... (1750); Kongl. Maj:ts Förklaring... (1757); Kongl. Swea Håf-Rätts Bref... (1757); Gratzer (2008). For more general contributions, see Shaiman (1960); Coleman (1974); Markham (1995); Finn (2003), part II; Vause (2016).
29 Schnabel and Shin (2001/2003); Hayen and Nyberg (2017), p. 38.
30 Kongl. Maj:ts Nådige Förordning... (1766); Kongl. Maj:ts nådiga Förklaring... (1767); Kongl. Maj:ts förnyade Stadga... (1773).
31 Kongl. Maj:ts förnyade Stadga... (1773); Agge (1934), pp. 916–917.
32 SSA, Stockholms magistrat och rådhusrätt, C5a, Konkursdiarier, vol. 1.
33 See for example Hayen (2017), pp. 141–155.
34 Kongl. Swea Hofrätts Bref... (1768); Agge (1934), p. 926.
35 Broomé (1888), pp. 10–19; Inger (1986), p. 221.
36 Bihang till Riks-Ståndens Protokoll vid urtima Riksdagen... (1817–1818), no. 18, p. 179f.
37 Tuula (2001), p. 288f.
38 Konkurstillsynsutredningen (2000), p. 119.
39 Tuula (2001), p. 292; Leijonhufvud (1991).
40 Adamson (1966), p. 158.
41 Bihang till Riks-Ståndens Protokoll vid urtima Riksdagen... (1817–1818), no. 18, p. 221f.
42 Brottsbalken den 21 December 1962... (1963).
43 Leijonhufvud (1991).
44 Nordisk Retsencyklopaedi (1878–1890), vol. 4, p. 202.
45 Underdånigt Betänkande till Kongl. Maj:t... (1853).
46 Konkurstillsynsutredningen (2000), p. 120.
47 Adamson (1966), p. 158.
48 Albinsson Bruhner (2004).
49 The breakthrough of the corporation form in Sweden was a time-consuming process, which can be said to have taken place over a period of at least 80 years. The actual breakthrough phase happened in the period 1895–1920. Broberg (2006).
50 Sallila (2016), pp. 87–89.

4 Bankruptcies in Sweden, 1774–1849

Causes and structural differences

Marcus Box, Karl Gratzer and Xiang Lin

Introduction

In 1771, the first Swedish academic thesis on bankruptcy and insolvency was defended by Carl Bergström at Uppsala University.[1] In this and other contemporary Swedish publications on the topic, shortcomings in the debtor's character including gambling, dishonesty, fraudulent behaviour and a disposition for speculation were mentioned as major causes for bankruptcies. A lifestyle with a propensity for consumption of foreign fashion and luxury items was also mentioned as a way that would lead to bankruptcy.[2] This was reflected in contemporary legislation and in economic doctrines. Mainstream explanations for why bankruptcies occurred, as recounted both in the theoretical literature, but also for example in theatrical plays in the eighteenth and nineteenth centuries were generally simplified and sometimes unrealistic. Only rarely was any distinction made between the person and the failure of the enterprise.[3]

Where did these harsh judgements of the insolvent debtor and the opinions about the factors causing bankruptcy stem from? One of the roots was unquestionably the old Roman bankruptcy law. Another later impact was the medieval Italian bankruptcy legislation, which overall was important for the design of the early modern bankruptcy institution and the view on the insolvent debtor. Baldus de Ubaldis, a medieval Italian lawyer, summarized the essence of the contemporary Italian bankruptcy law in the following statements: *fallitus, ergo fraudatur* (insolvent, thus, a swindler) and *fallity sunt infames* (insolvency means disgrace).[4]

The idea that a debtor also was a swindler, and should be severely punished, was spread by Italian merchants to, above all, France, Spain, England and Germany.[5] Similar explanations and judgements can be found in many subsequent writings. This includes early modern economists, like Adam Smith, the founder of classical economic thought. In Smith's economic system, governed by the invisible hand and populated by people who do not make any mistakes, bankruptcies were rare events. Not surprisingly, Smith regarded almost all bankruptcies as fraudulent or as a result of criminal speculation or smuggling. Therefore, bankruptcy was a crime that should receive the most severe punishment.[6]

Similar explanations can also be found in the French literature.[7] One of the few exceptions to this mainstream view was Anders Berch, the father of Swedish economic science, who published a textbook on the subject in 1747. In it Berch presented the first embryo to an explanation of bankruptcies on a meso level, when suggesting that most bankruptcies were caused by smuggling. Items imported without tax could result in the insolvency of entire industries, which in his mind was one of the most unsympathetic crimes.[8]

The moralising causal explanation for bankruptcy can be questioned from a social science research perspective. Based on modern literature, we can see many reasons for why a trader, shopkeeper or an artisan had to file for bankruptcy.[9] Very simplified, a distinction can be made between two different types of explanations. An *individual-centred* explanation focuses on the shopkeeper, the artisan, the entrepreneur or the management. In this scenario a business failure such as bankruptcy is regarded as caused by a lack of expertise and experience. The other explanation is *environment-related* and suggests that bankruptcies are triggered by changes in the firm's environment that could have consequences for its survival chances. Examples include the institutional and legal environment, such as insolvency laws, or the behaviours and actions of other actors such as creditors and banks. Other examples include variations in exchange rates, or business cycle fluctuations, and external economic shocks to the economy such as macroeconomic and financial crises. An economic shock is an event that occurs outside of an economy and produces significant change within an economy. Economic shocks are random and unpredictable events that have widespread and lasting effects on the economy and can be the root of unemployment, inflation and recessions; they can impact the economy in a positive or negative way through either the supply or the demand side.[10]

This is not the place to discuss the various personal characteristics, traits and motivations that have been highlighted in earlier studies. It is sufficient to conclude that this type of individual-oriented perspective has been exceptionally popular in the literature on business studies in general. While important, this is not the only perspective that influences entrepreneurial and business behaviour.[11] In reality, it is likely that different individual-centred and environment-related explanations overlap and reinforce each other in a complicated way. It is often difficult to investigate this relationship quantitatively, even when using modern, contemporary data and statistics. In historical studies, the task is often even more difficult, if not impossible, due to limited access to robust data.

Our impression is that historical research on bankruptcies has been dominated by different case study approaches. Many of these, often descriptive studies, have provided valuable insights into creditors' and debtors' relations and behaviours as well as into the consequences these insolvencies had on the actors involved. In-depth studies have also created fascinating knowledge about how companies were organized and how the credit networks and commercial structures functioned.[12] They have also contributed with new information about the importance of personal credit networks and trust in daily business life.[13]

Notwithstanding all of these benefits, the case study approach is often not suitable for generalization. We rarely know if something is unique and time-bound or general and possibly formulaic behaviour. The results from case studies are sometimes anecdotal and most often not statistically generalizable. The micro-oriented explanation of the causes for bankruptcies can be reductionist. If the unit of investigation is a single case (a firm or an entrepreneur), the explanation of the failure becomes oriented towards the company or the individual. If we in this way assume that it was the entrepreneur's personal traits and skills that caused the commercial failure, economic mechanisms related to the environment or other hidden but possibly important variables are easily overlooked.

In the following, we will show how quantitative methods can complement qualitative case-study approaches. The methodological aim of the study is twofold. First, we try to break with the traditionally descriptive nature of bankruptcies in historical research. We intend to provide a more analytically oriented approach where hidden and hitherto unknown – primarily structural or macroeconomic – causes behind the development of bankruptcies can be made visible.[14] Second, as a result, we try to demonstrate the potential of alternative research approaches by constructing longitudinal aggregate series on bankruptcies and on economic and demographic macro data. We view the approach as a valuable complement to an essentially micro-oriented descriptive research approach.

The empirical aim of the study is to highlight and make visible the connection between demographic and macroeconomic change and the occurrence of bankruptcies. The investigation is based on the development in the two Swedish towns Gothenburg and Stockholm between 1774 and 1849.[15] We will explore if there was any correlation between the volume and the variation in bankruptcies, on one hand, and variations in economic output, demographic changes and foreign trade, on the other. Did changing macroeconomic conditions affect the bankruptcy rate in different towns in a similar way or was the effect different? By studying bankruptcies on a macro-level, we will create a fundament for other researchers that can be used to make future international comparisons.

The economic development of Stockholm and Gothenburg

Sweden was a poor country in the early modern period with living standards well below some continental and Southern European countries. Real income per capita stagnated for most of the period from the sixteenth to the early nineteenth century.[16] At the beginning of the nineteenth century, it was still an overwhelmingly rural and agrarian economy, with a level of urbanization that was less than 10 per cent. Due to low income levels when compared to other European nations, it would benefit from its relative backwardness. Rapid industrialization and economic development in the late nineteenth century enabled a period of rapid catch-up growth relative to Western Europe.[17]

The Swedish capital Stockholm had growth problems in the century after 1750 and has been described as a stagnating city during this period. There were several reasons for this. Only a small number of businesses experienced successful growth and developed their workforces. As an important centre of trade, Stockholm still declined in a relative way rather than in absolute terms. Within the handicraft sector in the 1760s, profits were depressed during the latter part of the eighteenth century. No substantial expansion took place during the first part of the nineteenth century. Real wages of labourers were falling, reaching a bottom around 1800. By 1850, the purchasing power of a worker was approximately the same as in 1760. It was not until the 1840s when there were tendencies for improving economic development. The total number of artisans was reduced during the second half of the eighteenth century as low profits forced many into bankruptcy.[18]

Gothenburg and the western coastal region in general experienced a massive herring fishing boom after 1750, with a peak between 1780 and 1800, which in scale and economic significance, was unparalleled.[19] As a result the part of the population who drew their livelihood from fishing and fish processing experienced a few decades of rare brilliant economic conditions. From the early 1800s there were increasingly poor catches of herring. When the herring shoals suddenly disappeared in 1809, the coastal region around Gothenburg was hit by an economic crisis with catastrophic consequences. Large costly facilities were closed and equipment and tools were no longer needed. As a result, the coastal population decreased. In Gothenburg, traders were forced into bankruptcy. The capital they had invested in the fishing industry was lost, and they found it difficult to settle debt obligations or clear instalments.[20]

The impact of the crisis was offset by an upturn in trade and shipping caused by the Napoleonic Wars. The war between France and the United Kingdom created major problems for neutral trade. When Napoleon switched from direct combat measures to a trade war against the United Kingdom in 1806, Gothenburg was suddenly given a very favourable position, with a good part of the illegal trade to the British Isles channelled through the town.[21] Even after Sweden was forced to join the continental blockade, under the threat of being attacked, the United Kingdom continued to use Gothenburg for transit trade. The British had the silent support of the Swedish authorities, and the business continued with false shipboard papers and large scale organized smuggling. The effects on the city were immediate. Growing demand for labour led to an increase in population, wages and also price increases. The defeat of Napoleon at Waterloo in 1815 put an end to the boom. Foreign businessmen left the city, some of the trading houses went bankrupt and the city became insolvent and indebted.[22] The flourishing trade atmosphere turned into economic crisis, and it would take several decades before the economy recovered.[23]

Bankruptcies and macroeconomic variations

Previous research on bankruptcies and macroeconomic variations has identified a connection between shifting business cycles and bankruptcies. Since long, it has been commonplace among economists to assume that business failures and bankruptcies are linked to the business cycle. Furthermore, financial variables – changes in interest rates, exchange rates and terms of trade – have been viewed as important causes of explanations for changes in the bankruptcy volume.[24] In times of crisis or recession, bankruptcies or the risk of bankruptcy tends to increase. In a booming economy, the opposite holds true. In a similar manner, economic historians have often discussed how the room of action for firms and business activities are affected by macroeconomic and cyclical contexts.[25]

In Sweden, economic crises due to overproduction began with the industrial revolution in the late 1800s. The history of the pre-capitalist business cycle instead has often been characterized as a history of trade crises.[26] These have been started and fuelled by over-speculation in certain areas, for example, commodity hoarding or commodity trading, corruption and consumption of luxury items.[27] The first international speculation crisis, the so-called Mississippi Bubble, occurred in 1720, as the final disaster of the mass speculation in France that followed in the wake of the founding of the French General Private Bank and the Mississippi Company three years before. The collapse of the bank and the company coincided with the British South Sea Bubble and plunged France and other European countries into a severe economic depression. The episode, which spread to several other countries, is considered a prototype of a pre-capitalist crisis.[28]

The substantial international ramifications of occurrences such as the Napoleonic Wars or the Mississippi Bubble suggest the wider impact of pre-capitalist trade crises.[29] In our analysis, this is considered an important factor, and we therefore assume that changes in Gross Domestic Product (GDP) per capita influence bankruptcy developments.[30]

A study of urban and economic development must be based on population changes. While population change is less than a perfect indicator of economic growth, it has still been used as a proxy variable in previous research. We follow this approach and assume a positive correlation between the two variables. Population growth will be assumed to correlate with good times as well as high-economic growth. We furthermore assume that a larger population creates more companies, which leads to more bankruptcies.[31]

We also assume a correlation between foreign exchange risk and bankruptcies. The possibility that the market value of assets vary due to changes in the value of a nation's currency is the foreign exchange rate risk that the residents of that country face because of variations in the currency. For instance, if a Swedish merchant house had many loans denominated in the Hamburg *Reichstaler* but earned nearly all revenues in the Swedish *riksdaler* from sales within Sweden, a weakening of the Swedish currency meant that

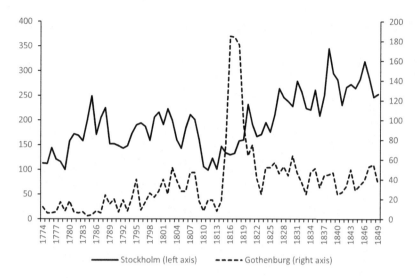

Figure 4.1 Number of bankruptcies in Stockholm and Gothenburg, 1774–1849

the company would have to allocate a larger portion of its earnings to make the same loan payments as before. A fall in the value of the *riksdaler* would increase the operating costs of the firm, thereby reducing its profitability and raising the likelihood of bankruptcy.

Our investigation is based on published and unpublished records about bankruptcies sourced from two different archives. For Gothenburg we have excerpted data from the Register of Bankruptcy Proceedings at the Gothenburg County Archives.[32] For Stockholm, we have used bankruptcy registers at the Stockholm Magistrate's Archives at the Stockholm City Archives.[33] Figure 4.1 shows the number of bankruptcies in Stockholm and Gothenburg between 1774 and 1849. As can be observed there were substantial variations in the bankruptcy frequency in both Stockholm and Gothenburg over time. In addition, Stockholm generally had substantially higher levels of bankruptcies during the observation period. One very clear exception is the very high bankruptcy levels in Gothenburg between 1816 and 1820.

Furthermore, Figure 4.2 describes the bankruptcies per 10,000 inhabitants in the two towns. It can be noted that, even after taking population size into consideration (the population of Stockholm was approximately four times higher than Gothenburg's throughout the period of study), the bankruptcy rate was generally systematically higher in Stockholm. Again, one obvious exception from the rule is Gothenburg's rising bankruptcy rates during the second decade of the nineteenth century. Overall, the variations in the bankruptcy rates of Stockholm and Gothenburg during the period of observation (see Figures 4.1 and 4.2) suggest that they could be caused by cyclical factors and structural changes in the economy.

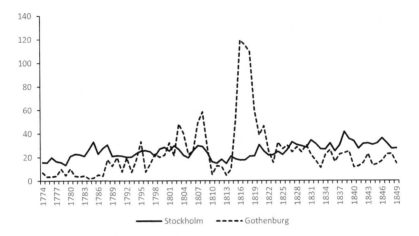

Figure 4.2 Bankruptcies per 10,000 inhabitants in Stockholm and Gothenburg, 1774–1849

Statistical investigation

In the statistical analysis, we use statistics on population and changes in the macro economy, here measured as the GDP (fixed prices; 2000 = 100). However, there are problems and pitfalls when applying our method on an early modern material. Often, deeper longitudinal statistical analyses of phenomena in larger populations are made impossible by problems with the statistical sources. This also holds true for early modern Sweden. Available data on population development and economic cycles for this mainly pre-quantitative area are unfortunately deficient.[34] Initially, we attempted to employ changes in exchange rates as an explanation for bankruptcies. However, there is a lack of unbroken series of exchange rates for the period of investigation.[35]

Therefore, as a proxy, we have instead used statistics on the number of outgoing ships from Swedish ports. Sweden's location, as a peninsula in Northern Europe, meant that virtually all imports and exports were shipborne. We have assumed that the number of outgoing ships could serve as an indicator of the changing conditions for foreign trade.[36] Even this attempt was blocked by the absence of reliable data. It was not possible to find robust data either on outgoing ships for both towns or for the volume and the value of foreign trade. The value of Sweden's export for the last four years of the 1750s has been estimated at approximately 63 million dollars silver money. At the same time, imports, according to official information, amounted to about 51 million dollars silver money. Because of various types of tax fraud, unpaid fees and smuggling, in reality, the real value of the foreign trade was likely twice that high.[37] Here, we only highlight these weaknesses but do

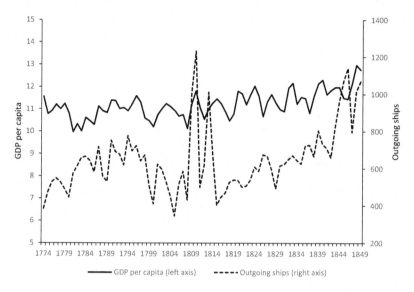

Figure 4.3 GDP per capita in Sweden and number of outgoing ships from Gothenburg

not attempt to compensate for the potential inaccuracies. Reliable data was only available for outgoing ships from Gothenburg (see Figure 4.3).

The dependent variable in our study is represented by the number of bankruptcies in a certain year. In order to compare the relative numbers of bankruptcies in Gothenburg and Stockholm we have created a new variable: *bankruptcies per capita*. While the population for Sweden and Stockholm can be determined with some accuracy for the period, data that can be used to determine the size of the population in Gothenburg on an annual basis is lacking.[38] We know that the population of Gothenburg during the late eighteenth century grew at roughly the same rate as in other Swedish cities.[39] Some general trends for the following period can also be established. The population reached around 17,000 people at the turn of the 1800s. It continued to grow and by 1815 had reached 23,000. This was followed by a long period of stagnation. In 1840 the population was estimated at about 25,000.[40] To overcome the lack of more detailed data, we have used statistical techniques for estimating the population in Gothenburg.[41]

Estimation method and results

In estimating our assumptions we employ a so-called truncated negative binomial regression model.[42] The analysis of bankruptcies in Stockholm and Gothenburg is carried out by regressions with GDP per capita, the population in Sweden and the populations in the two towns, as independent variables.

The result for Stockholm is reported in the right hand column of Table 4.1. For the countrywide population, the probability values are not consistent.[43] This leads us to conclude that changes in the Swedish population as a whole had no impact on bankruptcy rates in Stockholm. On the other hand, the probability values associated with the population in Stockholm indicates significance: an increase of the population with 1,000 persons – a not uncommon figure in Stockholm during the period of investigation – was associated with a 3.3 per cent increase of the bankruptcy rate.[44] When we consider the statistically significant (logarithm of the) GDP variable, the impact from the macro economy becomes obvious. Specifically, a 1 per cent fall in economic growth would increase the bankruptcy rate by 0.71 per cent, and a fall in GDP by 5 per cent would thus lead to an increase in the volume of bankruptcies by around 3.5 per cent.[45] Overall, our result for Stockholm indicates that demographic changes and cyclical variations in the macro economy represent plausible explanations for changes in the volume of bankruptcies in Stockholm during the period of investigation.

The story is quite different for Gothenburg. Based on the same independent variables, the estimation returns very low explanatory values

Table 4.1 Statistical results

	Gothenburg	Stockholm
Population in Sweden	1.003	1.000
p-value	0.000***	0.031**
p-value (robust)	0.000***	0.091*
p-value (bootstrap)	0.000***	0.119
Population Gothenburg	0.909	
p-value	0.001***	
p-value (robust)	0.000***	
p-value (bootstrap)	0.000***	
Population Stockholm		1.033
p-value		0.000***
p-value (robust)		0.001***
p-value (bootstrap)		0.017**
lnGDP	0.376	0.294
p-value	0.648	0.018**
p-value (robust)	0.648	0.016**
p-value (bootstrap)	0.664	0.038**
Outgoing ships from Gothenburg	0.997	
p-value	0.000***	
p-value (robust)	0.000***	
p-value (bootstrap)	0.000***	
LR test of alpha = 0	Chibar2(01)=907***	Chibar2(01)=454***
Pseudo-R^2	0.08	0.09

*, ** and *** represent the significance of corresponding coefficients (not the p-values) at the 10%, 5% and 1% levels, respectively.

when compared to Stockholm.[46] Thus, a new explanatory variable, the number of outgoing ships from Gothenburg, is introduced. The explained variance significantly improves the explained power of the model which reaches a level similar to that for Stockholm (pseudo-R^2 in the last row in Table 4.1).

Variations in GDP did not significantly impact the number of bankruptcies in Gothenburg. However, overall population growth in Sweden together with other variables had a statistically significant impact. The results show that changes in the general Swedish population affected bankruptcies in Gothenburg: a 1,000 increase in the population led to a 0.3 per cent rise in the bankruptcy rate. Consequently, an increase by 10,000 inhabitants in Sweden resulted in a 3 per cent increase in bankruptcies. Even if there were exceptions during the period of observation and several times even negative population growth, the Swedish population generally increased with approximately 10,000–40,000 people per year. Consequently, and just like in the case of Stockholm, population changes had substantial effects on the volume of bankruptcies in Gothenburg.

However, an increase in the population in Gothenburg had opposite effects, since it would reduce the number of bankruptcies (as an example, an increase by 100 people reduced the bankruptcy rate by a little less than 1.0 per cent). This contradicts our previous assumptions and we have currently no explanation for this relationship. However, at the same time, the regression results show that an increase with one (1) outgoing ship would reduce the bankruptcy rate in Gothenburg by 0.3 per cent. Since there could be a considerable variation in the number of outgoing ships even from one year to another, sometimes differing by several hundred (Figure 4.3), this implies that an increase by 100 ships decreased the bankruptcy rate by no less 30 per cent. The result therefore suggests that increasing exports substantially affected the number of bankruptcies in Gothenburg during the period of investigation; consequently, falling exports led to considerable intensifications in the bankruptcy rate.[47]

Overall, the regression model in our statistical investigation can explain a little less than 10 per cent of the total variation in bankruptcies (see Table 4.1, pseudo-R^2 for Gothenburg = 0.08; for Stockholm = 0.09). This means that more than 90 per cent of the changes in bankruptcies in the two towns during the period of investigation are unexplained in the formal analysis. The model is therefore not suitable for predictive purposes. Rather, it should be viewed as a quantitative exercise that provides a qualitative picture of the notion that demographic and economic changes were important in explaining bankruptcy.

Conclusions

Using historical statistics, we have attempted to make visible structural- and macro-explanations for bankruptcies, using Sweden and the two largest towns Stockholm and Gothenburg as a case in point. Our overall aim has been to show how quantitative methods can complement qualitative case-study approaches and how they can challenge individual-oriented explanations. Several previous attempts in historical studies to explain the causes of bankruptcies have been descriptive and often qualitative. Case studies can provide us with in-depth knowledge on several issues but they are less suitable for generalisation.

In the eighteenth and nineteenth centuries, a predisposition for consumption of foreign fashion and luxury items was mentioned as causing bankruptcy, along with gambling, speculation and dishonesty. In essence, shortcomings of the debtor's character and moral were held as main causes, a view also reflected in the legislation and the economic doctrines of that day and age. However, despite that the actions and decisions of individuals, managers and owners are of great importance, we have here suggested that it is more likely that individual-centred and structural- and macro-explanations are complementary. Indeed, our results suggest that both demographic and macroeconomic factors are plausible, partial explanations for changes in bankruptcy rates.

Our statistical investigation has shown that population changes and demography represent an important factor despite the fact that there were some inconsistencies in the empirical results. We assumed that population growth can be linked to macroeconomic growth. An increasing population will generally lead to intensifications in economic activity in general, including the creation of more businesses – potentially increasing the bankruptcy rate. Overall, we can establish that changes in the population at both national and local levels affected bankruptcy rates in Stockholm and Gothenburg during the period of investigation. However, nationwide population changes impacted Gothenburg, but not Stockholm. Changes in the local population also gave dissimilar results: it increased the bankruptcy rates in Stockholm but reduced those of Gothenburg. These discrepancies clearly need more investigation. We believe that shortcomings in demographic data may explain these results.

Another main macro-explanation used in our study was economic growth and economic cycles. Business cycle changes and economic growth had a clear impact on bankruptcies in Stockholm. This implies that recessions would lead to higher bankruptcy rates and to lower rates during economic upturns. Again, we had some inconsistent results since this relationship could not be established for Gothenburg. However, we found very distinct relationships between Gothenburg's bankruptcy rate and cycles in foreign trade – falling exports substantially increased the number of bankruptcies while bankruptcies would fall during upturns in exports.

Two other structural differences between the two towns are also noticeable. These differences were not made visible in the statistical investigation but can be observed by graphic inspection. To begin with – and with notable exceptions that are discussed below – the per capita bankruptcy rate in Stockholm was systematically higher throughout the period of investigation.

This difference may reflect a more intense economic and business activity in the capital of Sweden compared to other parts of the country.

Second, Gothenburg is a distinct and interesting example of a town affected by external economic shocks during the period. The shocks do not seem to have left any traces on the local bankruptcy rates in Stockholm, suggesting that the economy in Gothenburg was more connected to international networks than Stockholm's. For Gothenburg we could observe two supply-side chocks reflected in high bankruptcy rates 1803–1808 and 1816–1820. Intensifications during the period 1803–1808 have a plausible explanation in the increasingly poor catches of herring and the complete cessation of the herring trade in 1809. Gothenburg's position as a hub in the illicit trade to the British Isles, however, probably had a positive effect in counteracting the negative effect from the fall of the herring fishing; the relatively low bankruptcy rates between 1809 and 1814 indicate such a positive external shock. This profitable business had served as a growth engine for the entire city and may be assumed to lower the bankruptcy frequencies. Another and even more noticeable event occurred outside the Swedish economy and produced significant change in the business conditions of Gothenburg: the end of the Napoleonic war in 1815. During the two subsequent years (1816 and 1817), the bankruptcy rate in Gothenburg was exceptionally high. Thus, compared to the number of bankruptcies caused by, for instance, long term changes in population, the effect from external shocks on Gothenburg's bankruptcies appears to have been much more significant.

The overall approach in this study, using quantitative data and techniques on long term changes in bankruptcies, cannot solve all problems in the search for explanations. We believe that lack of historical data and of consistent series on economic variables may explain sometimes unpredictable results in historical statistical investigations, including the present study. However, our method can raise interesting issues for in-depth micro-level studies, suggesting that there often may be several competing and complementary explanations to bankruptcies.

Notes

1 Bergström (1771). Our study forms part of the research project "Firm demography and entrepreneurship in Eastern and Central Europe and in the Baltic region", which has received financial support from Östersjöstiftelsen (The Foundation for Baltic and East European Studies).
2 Arnell (1825); Äldre Swenska och ännu gällande lagar... (1816); Orsaken til närwarande talrika... (1799).
3 Sylvan (1942).
4 Baldus de Ubaldis of Perugia (ca. 1327–1400) was one of the more famous lawyers in fourteenth-century Italy. Important bankruptcy statutes were created in Venice (1244, 1395 and 1415), Milan (1341), Florence (1415). For this process, see Alexander (1891).
5 Gratzer (2008); Hunt and Murray (1999).
6 Smith (1762–1763, 1776).

7 Reynard (2001).
8 Berch (1747).
9 Gösche (1985), p. 63ff.; Gratzer (1999); Balcaen and Ooghe (2006).
10 See, for example, Levy and Barniv (1987); Mikhed and Scholnick (2014).
11 Box (2008).
12 Nyberg (2006).
13 See, for example, Nyberg and Jakobsson (2013).
14 For early approaches, see Gratzer (1995); Gratzer and Box (2002); Box (2005).
15 The starting point of the investigation has been determined by the availability of reliable data.
16 Edvinsson (2005); Edvinsson et al. (2010).
17 Probst (2019).
18 Söderberg (1991).
19 Haneson and Rencke (1923).
20 Andersson et al. (1996).
21 This has been extensively described by Heckscher (1918).
22 The value of the Swedish *riksdaler* fell dramatically, which had effects both in Stockholm and Gothenburg. The Hamburg *Reichstaler*, which in 1807 was valued at 52–58 Swedish shillings, in 1815 had appreciated further and was valued at 116 shillings. This change forced 17 companies to file for bankruptcy in Stockholm. Börjeson (1932), p. 361.
23 Andersson and Sandberg (2018).
24 Everett and Watson (1998); Mikhed and Scholnick (2014); Lane and Schary (1989); Levy and Barniv (1987).
25 Box et al. (2016); Schön (1996).
26 Schumpeter (1939); Clausing (1959).
27 The crisis in Sweden 1763–1769 was for example reportedly caused by corruption and consumption of luxury items. Bergfalk (1859).
28 Kindleberger (1978).
29 Browning (2019).
30 Edvinsson et al. (2014).
31 Söderberg (1991).
32 Landsarkivet i Göteborg, Ämbetsarkivet, F13F, Övriga register, vol. 1, Register över konkursmål vid Göteborgs rådhusrätt 1700–1900.
33 SSA, Stockholms magistrat och rådhusrätt, F6a, Konkursdiarier.
34 Lilja (1992); Utterström (1954).
35 During our period of investigation, macro statistics are rare. The Royal Swedish Statistics Office was only established in 1858. The activity of the new authority was at the beginning mainly dominated by population statistics. Gradually, for example, agricultural statistics, economic and financial statistics, savings and bank statistics, poverty statistics and crime statistics were added, to name a few.
36 Data on shipping collected by Lind (1923).
37 Börjeson (1932); Åmark (1915).
38 The population statistics for Sweden and Stockholm has been taken from Befolkningsutvecklingen under 250 år. Historisk statistik för Sverige (1999). The population numbers for Gothenburg are available for every fifth year from 1795 onwards. Lilja (1996); Nilsson (1992).
39 Andersson et al. (1996).
40 Fällström (1974).
41 Specifically, we have used an interpolation method that assumes that the population in Gothenburg is a weighted sum of the population in Stockholm and in Sweden in total. The method handles years (observations) with missing data for the population in Gothenburg (as noted, population figures for Gothenburg

are available for every fifth year from 1795 onwards). In short, the interpolation specifies a linear function of the three populations. The determination of the coefficients in this function is based on the available observations of the population in Gothenburg. Missing values for Gothenburg are then loaded according to these estimated coefficients and the populations in Stockholm and in Sweden, respectively. The weights are rather stable at 1.03/1.02.

42 The bankruptcies in Stockholm and Gothenburg were not normally distributed. As commonly suggested, the Poisson and negative binomial distributions are better for describing this type of so-called count data. Zero-truncated negative binomial regression is used to model count data for which the value zero cannot occur and for which overdispersion exists. In statistics, overdispersion is the presence of greater variability in a data set than would be expected based on a given statistical model (e.g. the Bell curve). The negative binomial model is more general than the Poisson model since it does not require the mean and variance to be the same. The truncated binominal distribution model can rule out some impossible values. For instance, we can skip the value of 0, since there were always bankruptcies in Stockholm and Gothenburg in our sample. For the estimation, the truncated points are chosen as the minimum value in the observations. For Gothenburg, we truncate at 2 since the minimum bankruptcy count is 3. For Stockholm, the truncated point is at 98, since the minimum bankruptcy count is 99.

43 The coefficient is significant according to normal and robust (10%) but not significant with bootstrapped standard deviation.

44 The significance is 5 per cent according to the bootstrapped standard deviation.

45 ((1–0.294) * 100% * 0.01)

46 $R^2 = 0.03$.

47 We also pay attention to the specification tests in Table 4.1, since the Poisson distribution is a special case of negative binominal distribution. This is tested by the null of $\alpha = 0$. If the LR test rejects $\alpha = 0$, the negative binominal count data model is an appropriate model to be employed. The results reported by the LR test of *α = 0 show clearly that the null of α = 0* can be rejected. Thus we have employed the appropriate model in this study.

Part III

Credit and bankruptcies in the fashion and luxury trades in Sweden, 1730–1850

5 The institutional setting of the luxury trades in eighteenth and early nineteenth-century Stockholm

Klas Nyberg

Introduction

The foundation of Stockholm and its rise from the twelfth century onwards can be explained from a financial as well as defence-related point of view. As a transit point for copper and iron produced in mid-central Sweden it emerged as an economic hub that attracted German merchants in the late Middle Ages. By the middle of the seventeenth century, Sweden was Europe's leading exporter of copper, and later, during the eighteenth century, of high quality wrought iron. Most of this export was channelled through Stockholm. Domestically, the city belonged to a small group of staple cities with exclusive rights to conduct international trade. In this group it held a special position as an obligatory port of call for all goods arriving from and destined for harbours in northern Sweden.[1]

Stockholm was not just a major port and commercial city, however. It was the Swedish capital, where the elites convened, where the Diet met and where economic policies were shaped in central government offices. As the capital, Stockholm was also the national and political centre for Sweden-Finland, the Swedish Baltic provinces, as well as Swedish Pomerania, an area of control that encouraged major European merchants to establish themselves permanently in the city during the early modern period.[2]

The substantial immigration of merchants, manufacturers and artisans helped to bring the isolated Swedish economy into a wider European economic context. Foreign experts contributed with knowledge and capital: financial networks were established between Stockholm and the leading financial capitals in Europe, including Amsterdam and London. By the eighteenth century, the formerly exclusive Lutheran Swedish society was also inhabited by French-Reformed, Dutch-Reformed and non-Lutheran Germans. Following further reforms in 1781 and 1782, both Jews and Catholics were allowed to settle and practice their religion, albeit with restrictions.[3]

The concentration of immigrants in commercial metropolises was not unique to Stockholm. It has long been known that colonies of immigrant merchants played a central role in the creation of dynamic trading areas. In many different places in the European periphery, in Southeast Asia and in North America, mercantile colonies were established as outposts to commercial metropoles in Europe and Asia. They were by no means isolated settlements but were connected through transnational kinship and credit networks. In Stockholm, foreign merchants with a background in the larger commercial metropoles in the European centre not only provided network links and connections but also crucially credited both to the monarchy and the nobility, as well as to manufacturers and other business operators.[4]

Already from the 1740s onwards, when government reforms were still lacking, the Stockholm Wholesale Merchant Association took upon itself to create a financial infrastructure, with solutions that made it easier for the private credit market to function. The society operated on an intermediate level between public and private that preceded the specialized capital markets of the late nineteenth century. As will be shown, wholesale merchants in Stockholm would continue to function as driving forces in larger economic-political shifts by shaping state fiscal regulations and institutions during subsequent periods of social development in the eighteenth and early nineteenth century.[5]

The trading colonies that emerged in different parts of the world were not only relevant for financial aspects of modernization such as knowledge of financial assignments, promissory notes, information on shipping, stock market quotes and exchange rates. The cross-fertilization of language, ethnicity and religion, legal traditions and innovations has led the historian Kapil Raj to speak of mercantile "contact zones" where specialist knowledge contributed to the creation of a new kind of "scientific knowledge construction" that incorporated: "[...] natural history, terrestrial surveying, map-making, law, linguistics, and public administration", but also "[...] the grouping of modern legal, political and administrative practices [...]".[6] The creation of this new science-based knowledge was helped by bureaucratic practices developed by state and municipal officials.

Raj focuses on the relationship between Europe and South Asia and in particular Calcutta, from 1773 the capital in British India, which became a centre for education, industrial know-how, science, culture and politics. I would argue that Stockholm, with its multicultural character and location on the periphery of Europe, in a derivate sense also can be viewed as a "contact zone" during the early modern period.

As mentioned, Stockholm's prominent immigrant merchants led the way in the Swedish development, often as members of the Hat party, the political party which headed the restructuring of the Swedish economy in the period up until the 1760. Most of the merchants lived in close vicinity to each other in the centre of Stockholm, including at Skeppsbron, on the easternmost side of the Old Town, which during this period became a preferred place of residence for affluent merchants with an international background. Most of the major merchants were generalists who invested in various industries and undertakings. This included the export of traditional commodities such as iron, copper and tar. Imports however also increasingly started to include new consumables, including colonial and luxury goods, semi-finished goods, but also raw materials for manufactures. The most successful merchants operated their own ships and shipping companies and also sometimes functioned as private bankers. As mentioned, their loan operations played an important role in the development of the credit market, an input which happened both on their own accord, but also as part of their collective organization in the Wholesale Merchant Association.[7]

In the following, the role played by the leading Stockholm merchants as the reorganizers of the Swedish economic policy during the eighteenth century will be discussed in more depth. I will show how this group already at the start of the Age of Liberty (1718–1772) managed to implement an economic policy that was directly aligned with the interests of Stockholm-based exporters at the expense of importers, prominently including textile importing wholesalers. A particular focus will be on how this affected the institutional structure of the luxury and fashion textile industry and artisanal handicraft sector in Stockholm, including the production of silk and woollen fabrics, fine handicraft items and fashion products. In this respect, the chapter serves as an introductory background to the case studies that follow in Part III. A further aspect which will be added to the discussion relates to the long-term economic and demographic stagnation of Stockholm, a development that set in from the 1760s, and which would continue until the middle of the nineteenth century.[8] Most interesting in the present context is how this development should be viewed in relation to the various attempts to modernize the economy that took place during the same period, including the reformation of the bankruptcy institution.

The manufacture system and the Stockholm merchant community 1718–1800

Before the Age of Liberty, luxury and fashion goods, prominently including textiles, accessories, furnishings, tapestries and furniture fabrics were imported to Sweden by merchants who were exclusively active in this type of trade. During the latter part of the seventeenth century, the leading exporters co-existed with the specialist wholesale importers. With the emergence of increasingly mercantilist economic policies at the end of the century, the conditions for textile importers, in particular, were gradually undermined.[9]

The process came to a temporary halt for 20 years during the Great Northern War (1700–1721). After this, the same endeavours were resumed, and a group of officials and major wholesalers succeeded in enforcing an economic policy that gave Stockholm and the leading exporters an even more privileged special position at the expense of the importers. From 1727 onwards, an increasing number of import restrictions and sumptuary laws were introduced that in all essentials meant that the import of finished textiles became prohibited.

The demand for textiles was instead expected to be met by production in domestic manufactures. Initially, these could not compensate for the loss of imported textiles, however, neither in terms of scope nor quality. The former textile wholesalers were required to rethink or abandon their businesses. The long-term result was a significant decline in the importance of the group, who basically were called on to serve the manufactures as merchandise or textile retailers.[10] From having been at an income and wealth level equal to many exporters, the textile importers dropped down to the average level for all retailers.[11]

The death of King Charles XII (r. 1697–1718) and the fall of absolutism led to political conflicts and a deadlock that in hindsight can be viewed as the starting point for the construction of the foundations of a new economic policy. The stalemate fundamentally impacted the capital, where the major conflict was between state and municipal officials and the traditionally self-governing burghers. The conflict was intimately connected to the fact that the burghers alone paid municipal taxes, even though other groups including noblemen, high administrative officials and public servants and not least members of the military also formed part of the population.[12]

Overall, the position of the burghers was strengthened in relation to the municipal administration during the conflicts. At first, the municipal administration held sway. The local officials relied on the state support, which in Stockholm was personified by the Governor, the leading state-appointed official who oversaw the affairs in the city. The burghers for their part were represented by a body of 48 senior representatives. This body was weakened by the dominance of the municipal administration and was convened only on their initiative. In subsequent conflicts over the responsibility for the payment of various extraordinary expenses, decisions on administrative issues, control over appointments and other similar issues, the power of the administration was weakened, however, and the position of the burghers strengthened.[13]

In this power struggle between the state, the municipal officials and the burghers, a group of increasingly powerful merchants soon distinguished themselves from the other players. While state and municipal officials, burgher representatives and burgher guilds were bound by their respective institutions, the merchants operated more independently. They were still in contact with and were themselves also represented in some of the other groups. In connection to debates on financial issues, for example, the municipal administration customarily summoned not only the senior burgher representatives, but also commonly the merchants. The same was true when state economic policy guidelines were formulated in the Board of Commerce. In summary, the leading merchants were formally and informally represented in the most important administrative forums and political assemblies where the new economic policies were decided.

At the beginning of the Age of Liberty the merchant group was primarily made up of exporters active in the iron trade. At the time the group consisted of some 25 exporters, where 10 formed the top strata, and the rest were taxed at a slightly lower level. Many lived in houses at Skeppsbron, on the easternmost side of the Old Town, directly overlooking Stockholm harbour. As a result, the group as a whole has habitually been referred to as the Skeppsbro nobility. The term should not be interpreted literally, however, but used more as a symbolic term. There were in fact major exporters who resided elsewhere in the Old Town and on Norrmalm and Södermalm – the areas north and south of the city centre, respectively. Skeppsbron still was and continued to function as an important symbol, where the major trading firms were headquartered and where a lot of the actual logistics in the trade was conducted.[14]

The leading group also included importers, especially those who traded in textiles and consumer goods. While they primarily worked as wholesalers and distributors to smaller merchandise retailers, they often also combined this with private retail operations. As early as 1720 the textile importers began to be classified separately as silk and woollen merchants. While their income level was lower than that of the exporters, it was still comparatively high. When the leading exporters were taxed at 500 dollars silver money, or more, with exceptionally successful exporters paying as much as 1,200 dollars, the silk and woollen merchants paid between 100 and 300 dollars in taxes. The median income for all 173 traders in 1720 in comparison was only 60 dollars silver money.[15]

Next to their relatively free status and independent political manoeuvring, the leading merchants differentiated themselves from the municipal officials and other burghers in one further respect: when stalemates arose in debates or discussions, they initiated and pursued a common burgher standpoint against the municipal government. These standpoints were decidedly one-sided, however. The group typically favoured the export industry and manufactures, by advocating protectionism and other types of support. This was in stark contrast to ideas for free trade as pursued by retailers and artisans who in numerical terms made up the majority of the burgher community.[16]

The new economic policy favoured by the leading merchants was introduced over a longer period, culminating in 1739, with the publication of a manufacture ordinance, and the establishment of a hallmark court that comprehensively regulated the oversight over and the production in all types of domestic manufactures. This had been preceded by a number of decrees and regulations focused more specifically on the production of textiles, including a 1720 sumptuary law and, two years later, a regulatory precursor to the 1739 legislation.[17] As a further step in support of domestic manufacturing, at the end of the 1720s, a 5 per cent levy was introduced on more than 500 imports, all of which potentially could be manufactured in Sweden.[18]

The support of domestic manufacture production reached a high point in the 1739 decree. Not only was a decree from the 1660s that banned manufacturers from selling their products in their own retail outlets lifted. Manufacturers were also freed from a number of municipal levies that were otherwise mandatory for all burghers. The sale of imported goods was at the same time all but criminalized, and restrictions were introduced that stipulated that only merchandise retailers who had interests in domestic manufactures were allowed to import manufactured goods for sale on the domestic market.[19] The development can be considered as the prelude to major disputes between retail interests on one hand, and manufacturers on the other, that marked Stockholm during the first years of the 1740s.[20]

The one-sided focus on the major exporters of British and German origin in Swedish research has hidden the fact that the eighteenth-century merchant community in Stockholm consisted of a myriad of traders and businesses of varying scale and life span. The official delineation between different professions conceals a more complicated interplay of activities carried out by individual merchants: financial transactions, traditional exports combined with imports and sometimes with involvement in handicraft or manufacturing.[21]

This clearly also applies to the wholesale and retail merchant trade sectors and to the distinction between wholesale merchants and merchandise retailers, who in reality represented many different kinds of business forms. Limited research has been conducted on the exact nature of the situation, however. Little is for example known about the majority of the smaller trading firms. Neither has a systematic investigation of the network relations between the different types of enterprises been undertaken. Even for the larger trading firms, the focus on the large exporters has meant that little attention has been paid to the trading firms that focused on imports. The same lack of knowledge applies to the various types of smaller and mid-size retailers.[22]

Only by acknowledging the variations, including the role played by marriages, social, ethnic, financial and genealogical networks as well as differences in scale and focus of businesses can the function of the Stockholm burgher community be understood properly. Research for example shows that an increasing number of wholesalers in Stockholm stayed unmarried during the eighteenth century and that this group had less financial success than their married counterparts. The development interestingly happened at the same time as the economic and demographic stagnation in the capital.[23]

Overall, about 100–130 wholesalers per year were active in Stockholm between 1740 and 1800. The sizes of their trading firms remained relatively constant over time. The group of small firms – here defined as operations that paid municipal taxes below the middle in the total distribution – ranged from 40 to 50. After the middle of the century, the number of large firms was around 30. The small firms accounted for a modest share of the financial result of the sector as a whole. Most often, their contribution was below 20 per cent. During the last quarter of the century, they only accounted for about a tenth of the combined municipal taxes levied on all wholesale merchants. The largest operations, conversely, increased their share over time. Until the middle of the century, they accounted for just over half of the ordinary taxation. At the turn of the nineteenth century, this share had risen substantially to 80 per cent.[24]

The gradual divergence and difference between wholesalers and merchandise and textile retailers that had begun in the earlier period was further strengthened with respect to average tax power between 1740 and 1800. Still, the merchandise retailers had the highest income level among the retailers, which remained stable during the second half of the eighteenth century. While grocers and, to a lesser extent, ironmongers strengthened their position they still earned 25 per cent less than the merchandise retailers, on average.[25]

Textile manufactures and the textile retail trade

While the importers lost in importance and the wholesale exporters gained in power in the commercial landscape that emerged after 1718, the most significant change came about following the rise of the manufacturers who benefited from the new manufacture system. Their rise and establishment in the Stockholm commercial landscape happened in particular after the manufacture-friendly Hat party came to power in 1738 and after the aforementioned publication of the manufacture ordinance and the establishment of a hallmark court the following year.

The steady establishment of new manufactures was only really challenged after the Diet 1765–1766, when a new political majority was established, and the role and impact of the manufacture-friendly leading merchants was curtailed. As mentioned, this also coincided with the start of a demographic and economic stagnation, which among other things also meant a long-term absolute decline of the manufacturing industry.[26]

Textile manufactures were the dominant type of manufactures in Stockholm. Within the textile sector, wool and silk manufacturing were the most important operations followed by cotton and linen as well as calico prints manufactures. To this could be added dyeing and shearing operations, both of which assisted the main textile industries.

In stark contrast to the situation when researching the various retail operations and trading firms during the period, annual factory statistics from 1739 to 1846 makes it possible to paint a highly detailed picture of the production in the domestic manufactures. This includes the number of practitioners and workers as well as the detailed production. The available statistics list the various manufactures according to the nature of their operation, in a way which also reflects the hallmarking procedure, which classified the fabrics and guaranteed the necessary qualities, as part of an inspection process during the production. The annual production figures refer to the quantity of fabric of a certain quality that a manufacturer had inspected. To this were added lists of workers that were compiled by the manufacturers themselves.[27]

Extensive research on the role and function of wool manufactures in Stockholm and Norrköping, a textile town some 130 km southwest of the capita, has shown that textile manufacturers increasingly formed part of a middle class after 1766. In wool-cloth manufactures – the most important of the textile sub-sectors – the incomes from the operations as well as the average household size decreased from the 1820, with fewer children and a lower age of the manufacturers themselves. The average size of the operations on the other hand increased during this latter period.[28] The importance of other cloth manufactures as investors and financiers also rose when compared to the late eighteenth century, when wholesalers provided funding for cloth manufacturers. A similar trend with an increasing reliance on relations within the closer group has been found for retail traders in Stockholm from the late eighteenth century, when compared to the beginning of the eighteenth century.[29]

Stockholm's textile trade was governed by unusually extensive regulations.[30] Only retail merchants and manufacturers were allowed to sell domestically produced fabrics. Retailers who sold textiles were traditionally referred to as silk and cloth retailers. They were not limited to sell those fabrics, however, but had the right to sell all kinds of Swedish-made textiles including knitted goods from manufactures and the handicraft production. To this could be added all legally allowed imported fabrics, as well as semi-finished fabrics and raw materials, as well as clothing made from the materials they were allowed to sell in raw and semi-finished forms.

When retail merchants sold the manufacture production, they were obliged to report annually to the Board Commerce about their transactions.[31] From 1739, the textile manufacturers were also allowed to sell their own production from factory outlets at the site of the actual manufacture. After 1830, they could pay a special fee for the right to sell freely, which included at periodic rural markets as well as by ways of wholesale and retail sales in urban shops.[32]

While silk and cloth retailers were allowed to sell handicraft textiles, this type of retail trade was more typically handled by linen retailers. While restricted to sell a more limited product range when compared to the former two categories, the linen retailers were more widely engaged in the manufacturing by providing processing and finishing work to the fabrics, yarn and clothes.[33]

When the fabrics were finished they were inspected and hallmarked by the hallmark court, meaning that they were ready for sale. The retailing did not take place on the court premises, however. Manufacture commissioners for example were obliged to sell in bulk from their dwellings and could only carry out retail operations if they had obtained special permission. Retailers were obliged to sell goods, in a room adjacent to the door of their houses, with specific prohibitions that the fabrics were not allowed to be stored in chambers or in the attic. The textile manufactures in Stockholm were also favoured over similar operations in other parts of Sweden. Only when a product that had been ordered from a manufacturer in Stockholm was not delivered on time would a manufacturer in another town be utilized.

In summary, the formal regulations regarding how textile trade was to be conducted were decidedly complex in Stockholm, even when compared to most of the rest of Sweden. The regulations were primarily set up based on the idea that textile manufacturers in Stockholm sold their products to textile retailers in Stockholm and only had their own stores at the production sites as a complement.

In reality, the situation was, as has been suggested, not without conflicts. The clashes between the major merchants and the broader burgher community that emerged in the early eighteenth century would continue for the rest of the century, with a peak during the 1760s. For the textile sector, the social differences and the conflict of interest between on one hand textile retailers and on the other wholesalers and textile manufacturers hampered textile production and trade. In addition, the detailed regulations that governed the way in which the different groups could operate led to an outdated and probably unwieldly arrangement for the sale of factory-made textiles. Both factors most likely added cumulatively to the general stagnation of the Stockholm economy.

The institutional structure of the luxury and fashion industry

The production of luxury items in Stockholm during the eighteenth century was carried out by highly specialized artisans. Besides employment in manufactures, artisans also worked as part of the guild system or were privately employed by aristocrats or the Royal court. Their work built on institutionalized concepts such as expertise and mastery as well as more artistic ideas of quality, creativity and technical proficiency. Both the guild-organized artisans and those who after 1739 worked in manufactures with royal privileges advanced their skill levels through structured education and rose in rank from apprentices to journeymen before becoming masters.

A careful review process was in place throughout the process, and masters and journeymen only achieved their eligibility after special exams. The review was handled by the guild to which the artisans were associated and which formed part of their social network. The principle of *numerus clausus* regulated the number of craftsmen who were allowed to operate in each guild, in a way which controlled the labour market in order to give each guild-member a reasonable income. Non-legitimate competitors were pursued mercilessly and brought to trial, whereas guilds with closely related activities were controlled to ensure that they did not infringe on the hard fought rights.

The guild system has often been interpreted in negative terms in economic-historical research, which has regarded it as a medieval institution that preserved an older situation. According to this viewpoint, guilds were detrimental to the expanding state by opposing technological change, new more efficient methods of manufacturing or generally the expansion of production.

With the rise of an industrial system that developed as part of commercial capitalism and the general growth of the early modern state two fundamental expansionist forces that challenged and opposed the guild system and burgher particularism were introduced. Another undermining force was the demographic transformation and the growth of the lower classes, including not least rural unskilled workers. When urbanization absorbed these individuals in the nineteenth century, a large industrial proletariat that challenged the highly regulated guild system emerged.

Guilds had been forced to deal with competition from artisans employed by monarchs and the aristocracy throughout the early modern period. In the seventeenth century a new challenge arose in the form of urban manufactures and new entrepreneurial employers. In Flemish textile cities, so-called "hallmark courts" were created as overseeing and controlling municipal institutions tasked with ensuring the quality of textile handicraft products. In the "halls" of the courts, mass-produced standardized quality textiles were measured and controlled in a process outside the jurisdiction of the existing guilds. The textiles had been manufactured as part of a cottage industry system, where entrepreneurial merchants engaged rural peasants to spin and weave, with the textiles eventually ending up in the export trade. When this production expanded from the end of the sixteenth and during the seventeenth century, the need for a special overseeing institution rose, and hallmarks courts were created to ease the burden on municipal governments.

In Stockholm, the hallmark court that was established in the eighteenth century was, as detailed above, initially focused mostly on the inspection of the textiles. The quality of the fabrics was examined already during different phases of the production process, crucially when it was taken out of the loom and later after dyeing. The actual product range was delimited to certain qualities and categories, regulated by the court according to international standards. In the event of technical errors during the manufacturing process or in the final product, fines were imposed by the court in accordance with a detailed technical specification. The complexity of the production was substantial. Upwards of a 100 different specified steps were necessary in the production of finer clothing qualities.[34]

Over time, the court would focus more on regulatory and judicial matters. The number of provisions relating to the control of the workers grew. The court also became increasingly integrated into the emerging state industrial policy, which saw it registering and keeping an account of the production in the manufactures and administering the operational concessions that had been awarded to the manufacturers.

While the establishment of hallmark courts had a background in a growing commercial capitalism and the international trade in textile products, where merchants organized standardized mass production, the hallmark courts in Sweden in their eighteenth-century manifestation, also integrated a range of luxury trades. This situation in the long run meant that the new institution also helped to break the role which guilds had over the basic conditions for work within the spheres of artistic and cultural production. It counteracted the guilds by abolishing social and cultural, formal and informal rights, including the idea of *numerus clausus*, whilst supporting technical change and product innovations.

The situation was not uniform. Some of the subdivisions in the luxury sector only partially fell under the jurisdiction of the hallmark court. Sculptors, gilders and painters at Rörstrand and Marieberg – the two major porcelain manufacturers in Stockholm – as well as tapestry producers answered to the hallmark court. The large and diverse group of painters on the other hand still belonged to the guild system after 1739. Furniture makers produced furniture and mirrors under the hallmark court jurisdiction. Carpenters were still organized as a separate guild, even when they were sufficiently specialized to be referred to as cabinetmakers. When under the jurisdiction of the hallmark court, they were sometimes instead referred to either as hallmark court masters or as ebonists.[35]

Other groups changed between the two jurisdictions including the small group of instrument-makers. Previous research has shown that even dyers and shearers, two of the key textile professions, vacillated in this way – contrary to the intentions of the hallmark court. This was further aggravated by the fact that many craftsmen often combined several kinds of professional skills in a way which required an association with both regulators.

A stagnating metropolis in a comparative perspective

In a now seminal work by Söderberg published in 1991, Stockholm was characterized as a stagnating metropolis between 1760 and 1850. The period was marked by a weak population growth and a declining population when compared to the Swedish population as a whole. The city was affected by a very high mortality rate that often exceeded the birth rate, declining nuptiality rates, the emergence of the so-called "Stockholm marriage" (cohabitation without marriage) and an increasing number of extra-marital children. In economic terms the capital lost its relative importance as a commercial city to Gothenburg, whereas its textile manufactures were out-competed by the manufactures in Norrköping, some 160 km southwest of the capital. Real wages and living standards fell, poverty increased and the income distribution was polarized.[36] The research that has been undertaken since has added to, but hardly changed the overall picture.[37]

Stockholm formed part of a group of European cities that also included Copenhagen and Amsterdam, which experienced weak population growth in the period between 1750 and 1850. The pattern was most evident in southern Europe, however, where a pervasive urban stagnation took place in Italian cities such as Genoa, Venice, Florence, Bologna and Rome as well as on the Iberian Peninsula with Granada and Lisbon. Conversely, there was a strong growth pattern in parts of Northwestern Europe where a number of British and Irish cities showed strong growth in the wake of the ongoing industrial revolution. A number of cities south of Amsterdam also showed strong growth after the turn of the nineteenth century. Paris and London as cosmopolitan centres also grew strongly. The same can be said of key metropoles in central and eastern Europe, including Moscow, St. Petersburg, Budapest, Bucharest, Warsaw, Prague, Berlin, Munich and Vienna.

In most of these latter cases the growth was more due to the emergence of a strong bureaucracy and military power than mercantile and industrial development, however.

The decline of Stockholm was apparent also in a Swedish perspective. One reason for this was a long-term decline in textile and luxury manufactures which were more or less completely eliminated in the first half of the eighteenth century. By the middle of the nineteenth century, only a few woollen and silk manufactures were left in the capital. Decreasing real wages for unskilled workers was so extreme that the purchasing power in 1850 corresponded to the situation in 1760 in absolute terms (see Figure 5.1).

As noted above, much of the production was taken over by Norrköping, a historically important industrial as well as textile-producing town in Sweden (see Figure 5.2). Two major reasons for the development can be offered. First, the organization of the industry in Norrköping made it easier to produce in a more flexible way and probably at lower unit costs than in Stockholm. The enterprises in Norrköping were a part of a wider infrastructure which enabled lower costs in many areas and a more effective market organization. The entrepreneurs in the town were also highly knowledgeable of local, regional and foreign markets.

The most important single factor to explain Norrköping's success, however, was the way in which the enterprises were operated and organized. While the majority of small-scale and medium-sized firms in Stockholm were run by former skilled workers, in Norrköping, the industry was dominated by textile retailers and dyers. When compared to older forms of industrial organization where marketing and selling were separated from the production, the textile entrepreneurs in Norrköping linked production knowledge and marketing knowledge. This happened in different ways. In

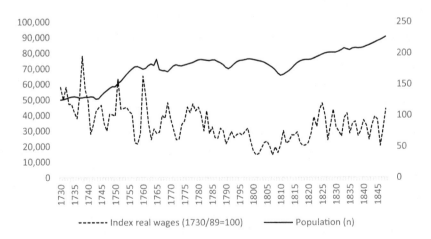

Figure 5.1 Population and real wages in Stockholm, 1730–1850

Source: Befolkningsutvecklingen under 250 år. Historisk statistik for Sverige (1999), for the population of Stockholm. Real wage index in Söderberg et al. (1991), pp. 88–91, appendix 2.

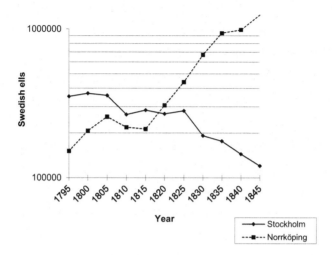

Figure 5.2 Production of woollen cloth in Norrköping and Stockholm, 1795–1845
(Swedish ells)
Source: Based on calculation by Persson (1993), p. 135, table 5.4.8.

some cases, knowledge of aspects that was lacking was acquired and was shared through kinship or family networks. In other cases, entrepreneurs themselves created joint enterprises which combined the production with dyeing, marketing and selling of textile goods.

In Stockholm, this was not the case. During the stagnation period, a shortage of capital, great fluctuations in demand and problems with financing meant that successful entrepreneurship in the textile luxury trades was dependent on organizational flexibility. Managers had to be able to minimize costs and to change and expand production in response to an increasing demand.

The importance of relative price shifts

The institutional development with the rise of specialized artisanal handicraft and luxury production that took place in Stockholm after the end of the Great Northern War, as described above, was not unique to Sweden. Indeed, capital cities with political centres, monarchs and an aristocratic presence often attracted clusters of specialized luxury artisans in the early modern period. The development in Stockholm however becomes interesting when considered in relation to the subsequent demographic and economic stagnation of the capital.

In textile and fashion research, analysis of relative prices plays an important role. The stagnation process in Stockholm has been interpreted against the background of the development of inflation in the eighteenth century, where the price of food increased more than the price of manufactured goods. The development is believed to have adversely affected profits in the manufacturing and handicraft sectors: "[...] we should expect growing problems with regard to industrial profits under the double pressure of rising labour costs (although nominal wage rises were usually small) and deteriorating terms of trade against the agrarian sector".[38]

From a demand point of view, however, the opposite was true. A falling price of, for example, luxury goods in relation to the price of food, promoted consumption. Luxury goods by the second half of the eighteenth century were demand by a relatively stable market segment. A similar development took place in Vienna in the first half of the nineteenth century. At that time, the production of cheaper textiles moved to the countryside in order to benefit from lower rural wages and other production costs. The silk industry, which was only gradually mechanized and built on skilled labour, just like in Stockholm, remained in the capital.[39]

The importance of relative price trends to explain the consumption of textiles and clothing has been noted for our own time, with the relative price of clothing for example halved in relation to the consumer price index during the post-war period in England and Sweden.[40]

Stockholm's textile luxury manufactures followed this trend during the eighteenth century. Fine woollen cloth, the most expensive and one of the most sought-after woollen qualities (see Figure 5.3) as well as droguet silk, which also belonged to the upper price segment (see Figure 5.4), decreased in relation to the price of rye in Stockholm (see Figure 5.5).

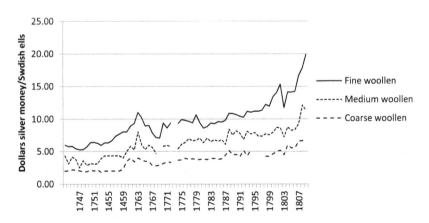

Figure 5.3 The unit price of the major woollen qualities, 1744–1810 (dollars silver money/Swedish ells)

Source: Unit prices have been calculated based on professor Eli. F. Heckscher's excerpt collection in RA, Eli Heckschers arkiv, Låda 24, Manufakturer i kuvert. For a detailed discussion about the material, Nyberg (2015), pp. 321–324; see Nyberg and Cizuk (2015), pp. 325–348.

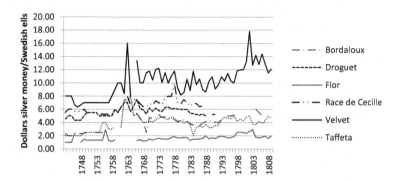

Figure 5.4 The unit price of droguet and five other silk qualities, 1748–1810 (dollars silver money/Swedish ells)

Source: See Figure 5.3.

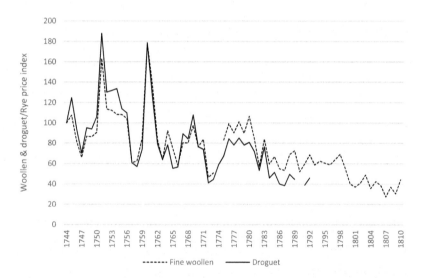

Figure 5.5 Relative prices of fine woollen cloth/ell and drouget silk price/ell to rye prices in Stockholm, 1744–1810 (1744 = 100)

Source: Stockholm rye prices in Söderberg et al. (1991), pp. 88–91 (SEK/hl), appendix 2. Woollen cloth and drouget silk prices based on RA, Kommerskollegium, Kammarkontoret, Da1, Årsberättelser fabriker serie 1, Generalsammandrag.

Conclusions

This chapter has dealt with the formation of the fashion industry and luxury crafts in Stockholm after the Great Northern War until the crises of the 1760s as a basic background to the case studies in Part III.

The chapter has detailed the long-term impact of the demographic and economic stagnation that characterized the city after this period from about 1760 to the mid-nineteenth century: weak population growth, a very high mortality rate, falling marriage rates and an increase in the number of extra-marital children. In financial terms, Stockholm lost ground as a trading town, and the textile manufacturing industry was de-industrialized. Rapidly declining real wages, declining living standards and increasing poverty were combined with an increasingly uneven income distribution. I have argued that Stockholm, with these characteristics, belonged to a group of cities that showed the weakest population growth in Europe, similar to cities in the Mediterranean, in Italy, southern Spain and Portugal.

The chapter has shown how luxury craftsmanship and the fashion industry were formed after the end of the Great Northern War. The conditions in the textile manufacturing industry indicate that the institutional structure of the fashion industry, which was characterized by social disagreements in the burgher community, weakened Stockholm's competitiveness during the stagnation. Perhaps was it the fixed, sluggish institutional barriers that determined who (exporters, importers, retailers, manufacturers, craftsmen) that were entitled to do what (manufacture, design, prepare, dye, sell, exchange, market) that formed part of the reason for the stagnation? The shift to the factory town of Norrköping during the early nineteenth century indicates that this was the case.

In the late seventeenth century the production process in the luxury textile trades was divided between carders, silk and woollen weavers, shearers, and silk and woollen dyers organised in traditional guilds. The whole working process seems to have been co-ordinated by traders. With the establishment of the hallmark court system in 1739, guilds were circumscribed and artisans were either subordinated to the hallmark courts or left answering to the guilds. In the textile trades, traders and clothiers were supposed to be the organizers of the production. Carders and shearers became workers whereas dyers remained craftsmen, but not organized in guilds, however.

In the late eighteenth century the Stockholm textile trade in a retrograde movement turned back towards the seventeenth-century situation, with three exceptions. First, the hallmark court system remained in place until 1846. Second, the larger textile companies declined in importance, even if they remained in business. Third, clothiers remained as the main organizers and co-ordinators of the woollen cloth working process. They bought imported wool from wholesale dealers and sold finished cloth to textile retailers.

On the whole, the textile industrial shift to Norrköping in 1825 meant a break with this old organization. Relatively large- and medium-sized firms increased their share of the total production volume between 1785 and 1835. In Norrköping, these enterprises were managed by tradesmen. The smaller firms were run by traditional clothiers. This development of spatially concentrated, more centralized and mechanized firms which integrated strategical resources and distribution channels was a new model distinct from the eighteenth-century crafts-based organization and the later nineteenth-century machine-based factory system.

Notes

1 Heckscher (1935–1949).
2 Kirby (1990).
3 Kirby (1995). Before 1774, Jews had not been allowed to live in Sweden at all, and the few Catholics had been forced to practice their religion in private or in the confinement of diplomatic residences.
4 Schunka (2005); Lesger and Noordegraaf (1995); Das Gupta (2001); Markovits (2000), p. 20f.; Chaudhury and Morineau (1999); Fleet (1999); Matthee (1999); Young (1995); Ojala (1997).
5 See Roberts (1986) for a general presentation of the Swedish Age of Liberty; Nyberg and Jakobsson (2012).
6 Raj (2007), p. 12; Raj (2011).
7 Nyberg (2001).
8 Söderberg et al. (1991).
9 Boëthius and Heckscher (1938).
10 The birth of the protectionist economic policy and the decline of textile imports are investigated in Aldman (2008). See also Ekegård (1924).
11 Boëthius (1943); Nyström (1955), bilaga VI.
12 Members of the military made up a third of the population of between 60,000 and 70,000 people. Boëthius (1943), p. 31.
13 Boëthius (1943), pp. 35f., 53. See Brolin (1953) for the crises of the 1760s and Nyström (1955), pp. 245–253. Riots in late eighteenth-century Stockholm are analysed in Berglund (2009), chs. 2–3.
14 Boëthius (1943), pp. 33–34.
15 Boëthius (1943), pp. 34–35.
16 Boëthius (1943), p. 145.
17 Boëthius (1943), p. 87; Kjellberg (1943), ch. V.
18 Boëthius (1943), p. 148.
19 Boëthius (1943), p. 190.
20 Boëthius (1943), pp. 212–214.
21 Samuelsson (1951), ch. IV; Gårdlund (1947), p. 85; Hasselberg (1998), ch. IV and p. 279f. See also Andersson (1988), chs. 4–5, for the merchants in Gothenburg.
22 Wottle (2000), ch. 1; Lindberg (2001).
23 Wottle (2000), pp. 226f.; Ågren (2007); Söderberg et al. (1991).
24 The analysis is presented in detail in Nyberg (2006), bilaga 3, pp. 364–366.
25 Nyström (1955), bilaga VI; Söderlund (1943), p. 335.
26 Söderberg et al. (1991), ch. 3.
27 Nyberg (1992), chs. 1 and 5 and app. I to chs. 5 and 6.
28 Persson (1993), pp. 83–84.
29 Wottle (2000), ch. 9.
30 See Dreutzer (1844), pp. 462–465, 594–595, for the legislation surrounding the textile retailers in Stockholm: "Borgmästare och råds handelsfördelning den 16/1 1641, art. I-VI"; "Kungl. reglemente för kramhandeln i Stockholm i Kungl. brev den 27/10 1749"; "Tilläggsförordningar 10/2 1805 och 5/6 1839".
31 First stipulated in "Kungl. brev 29/5 1750" and "Kommerskollegii cirkulär 6/2, 6/8 1786 och 10/4 1794", in Dreutzer (1844), p. 462. See also RA, Kommerskollegium, Advokatfiskalkontoret, Årsberättelser, "Upphandlade fabriksvaror" (ca. 1787–1820).

32 It was stipulated in the hallmark court legislation of 1739 and 1770. See also "Kungl. brev 15/7 och Kommerskollegii kungörelse 3/8 1830" in Dreutzer (1844), p. 291.
33 Lindberg (2001); Lindström (1923, 1929).
34 Berg (1967); Nyström (1955), p. 49f.
35 The latter was a direct translation of the French term *ébéniste*. Sylvén (1996), p. 14.
36 Söderberg et al. (1990).
37 Bladh (1991); Persson (1993); Nyberg (1999b); Wottle (2000); Lindberg (2001); Hayen (2007); Bladh (2008); Rasmussen (2010).
38 Söderberg et al. (1991), p. 18.
39 Schön (1979), ch. 11.
40 Jones (2006), p. 239; Gråbacke (2015), diagram 4.

6 Economic behaviour and social strategies in the Stockholm silk weaving industry, 1744–1831

Håkan Jakobsson

Introduction

The first silk weaving manufacture in Sweden was established in Stockholm in 1649 by an immigrant Dutch entrepreneur from Amsterdam called Jacob van Utenhoven.[1] It was not until after the Great Northern War (1700–1721), in the reign of King Fredrik I (r. 1720–1751) that an expansion of the industrial sector took place, however. This intensified in the late 1730s and in particular in the 1740s, when a number of enterprises were founded on the basis of new legislation and with the help of state subsidies. The growth continued, eventually stabilizing in the period 1780–1810, before starting a slow downward decline. As late as 1850, 18 silk weaving manufactures were still operating in Stockholm, not far from the number of plants that had existed some 50 years before. Shifting fashion trends and international competition pushed the industry into a quick decline thereafter. In 1869, only two silk weaving plants remained operational. All throughout, silk weaving and the broader industry that grew up around the processing, manufacturing and sale of products made from silk was an almost exclusively Stockholm-based phenomenon. Very few silk weaving enterprises existed outside of the capital.[2]

While textile historians have taken an interest in the textiles that were produced in the silk weaving manufactures, in history and economic history the industry has received little attention and has more often instead been downplayed as an anomaly with limited long-term relevance.[3] The latter opinion was prominently argued by Eli F. Heckscher who in the 1930s suggested that the entire range of manufactures that emerged in Sweden in the eighteenth century only really managed to develop because of extensive state aid, and in the end at best played a marginal role, in particular in relation to the nineteenth-century industrialization process. Most of the industries that shaped the Swedish industrial boom during this latter period, including industrial sawmill and ironworks, were not operations that had been included in the earlier manufacture system and hence logically did not directly build on this earlier development.[4]

Today, this picture has largely changed. Heckscher's opinion that manufactures existed only because of mercantilist policies and state aid was rejected by Per Nyström in 1955 specifically through an in-depth study of the Stockholm silk industry. While not arguing against the role played by the state in the establishment of silk manufactures, Nyström however stressed that the industry appears to have adapted and survived also after state subsidies were removed in the 1760s. He also broadened the outlook by placing the manufactures in a wider context, suggesting in particular the positive role played by merchant capitalists and thereby also pointing out the role of the manufacture system not as a forerunner to the industrial revolution but as a development in its own right.[5]

Nyström specifically suggested that the silk industry survived after it was forced to consolidate. It did so partly by adapting in size, but perhaps most prominently by changing to a more simplified production, and by producing less elaborate textiles with a much broader market potential, which quite possibly enabled it to survive long term.[6] The observation is clearly relevant when viewed in the perspective of the ideas about the growth of new luxury and shifting consumer behaviour. While Nyström far from conclusively established if it was fashion shifts that encouraged the manufacturers to change their production, or if it was the decision by the industry to develop new products that led to a rising market demand, the observations are nonetheless interesting. What it suggests at the very least is that the industry worked in conjunction with a consumer market and not simply survived because of extensive state financial support.

Although pointing out the flaws in Heckscher's argumentation, and highlighting the role played by fashion demands, Nyström was still influenced by Heckscher's basic problem formulation about the idea that the eighteenth-century manufactures were precursors to modern industries and hence should be supposed to play a role in relation to modernization. In addition, while arguing for a deeper look than purely statistical information about workers and operational enterprises, his observations about a shift from more elaborate and expensive to lighter and cheaper silk textiles were still made based on the same source material used by Heckscher, namely the well-preserved industrial statistics that exists for the period. While undeniably suitable for statistical analysis, in so far as to provide annual, relatively uniform information about the number of enterprises, workers and the production, this accessibility also informed his opinion of the character of the industrial landscape.

In this chapter I will discuss some aspects of the development of the silk weaving industry in Stockholm during the period 1744–1831. At first I will draw focus to its French roots, and the potential implications of this background, before turning to and specifically focusing on two periods which according to earlier research has constituted important shifts in the development of Swedish manufactures in general. The first period constitutes the years between roughly 1760 and 1767, when the manufacture system came under threat, in the wake of an international trade crisis (1763–1765) and after the suspension of the manufacture system and removal of state aid after the Diet of 1765–1766. The second period constitutes the years between 1805 and 1815 when Sweden in various guises became involved in the continental war against Napoleon. The unstable situation had great implications for the state of the Swedish economy, not least by cutting off access to a number of traditional markets, of crucial importance to the import and export trade.

When outlining the general development of the industry, I use the traditional industrial statistics utilized by both Heckscher and Nyström. In the end, I will however move away from this material and instead focus on the underlying social and economic structures that formed and shaped the industry. One of the aims of the investigation is to show how manufactures survived over time, but not as part of Heckscher's idea that they eventually would or could play an economic and industrial role in the nineteenth century. I instead focus on their specific role and function during the eighteenth century, both during periods of crisis, and in relation to bankruptcies, but also in general in relation to social and financial considerations. My conclusions will focus on the concepts of networks, cooperation and resilience, as three factors that marked the Swedish silk weaving industry during this period.

A growing industrial sector and its French roots

As mentioned, the first growth of the Swedish silk weaving industry from a peripheral industrial sector with almost no operators into something much more substantial happened in the period 1720–1750. Two major reasons for this development can be produced.

First, after the end of the Great Northern War and the abolition of absolutism, power shifted from the monarch to the nobility and to a political landscape centred on the Diet. During the ensuing, the so-called "Age of Liberty" party politics grew in importance, with two major political groupings, vying for power. After one of the political groupings, the manufacture-friendly Hat Party emerged victorious after the Diet of 1738–1739, new economic policies, favouring domestic production were quickly introduced. The support structure that was rolled out included a dedicated Manufacture Office, separate legislation, import barriers to promote domestic products, direct subsidies and cheap loans that would spur the desired growth of manufactures. The new policies were largely successful in the short term. It led to the mass establishment of all types of urban manufactures in the Swedish industrial landscape on an unprecedented scale.[7]

Second, and more specifically, the period saw the direct and indirect support of the domestic silk industry by leading groups in Swedish society. From 1744 onwards, and in particular, during the 1750s, this most notably included the culturally interested crown princess, from 1751, Queen Lovisa Ulrika, the consort of King Adolf Fredrik (r. 1751–1771).[8] In preparation for the coronation ceremony in 1751, the royal couple turned to the domestic silk weaving industry and commissioned Bartolomei Peyron, a Stockholm silk manufacturer of French extraction to weave the textiles for their coronation robes.[9] The appreciation of his work was apparent when he, a few years later, was awarded a newly designed gold medal for accomplishments

in support of Swedish manufactures. In 1755, a similar type of gold medal
was struck to be awarded to individuals who had supported the production of raw silk.[10] The idea was directly connected to an operation that the
Queen had set up close to Drottningholm Palace, west of Stockholm, a few
years earlier with the aim to secure a domestic supply of silk with the help of
a French silk cultivator.[11]

The use of French experts and artisans was no coincidence. French artistic and artisanal influences grew increasingly strong in Sweden from the
1730s onwards. The main reason for this was the need for expertise to help
in the reconstruction of Stockholm Castle.[12] The work was led by the Fran-
cophile and hat politician Carl Gustaf Tessin as head of state constructions
and more directly by his associate, the equally French-inspired architect
Carl Hårleman. Through the effort of the latter, a group of nine French
sculptors and painters were recruited in Paris and arrived in Stockholm in
the early 1730s. Over the years, the group would grow substantially, pav-
ing the way for an artistic influence with wide-ranging impact, effectively
introducing the ideas of rococo into Swedish architecture and a range of
associated cultural arts.[13]

Both Hårleman and Tessin had studied in and regularly visited France.
Tessin returned between 1739 and 1742 to serve as the Swedish ambassador
in Paris. During his stay, he immersed himself in French cultural life and set
off on a personal spending spree of enormous proportions, which saw him
purchasing paintings, sculptures, books and other luxury items.[14] After he
returned to Sweden, his successor on the ambassador post continued to function as a channel for cultural impulses through which the aristocracy could
secure luxury items.[15] When Tessin, after his return to Sweden was appointed
to serve as the Marshal of the Court to Adolf Fredrik and Lovisa Ulrika, his
contacts and style ideas gained further traction in the highest circles.

From the 1740s the silk industry was also directly supported by the Manufacture Office, which as indicated had been created a few years before to
keep oversight of Swedish manufactures. The body developed ideas on how
to attract foreign artisans and entrepreneurs to establish manufactures and
to organize the production. One of the key areas of interest was textile production, including the silk weaving industry. A key move took place in 1744
when Hårleman, who had taken over as head of state constructions after
Tessin, employed Jean Eric Rehn, a young Swedish draughtsman, who for
a couple of years had been studying engraving in France, to work in the
Manufacture Office.[16]

The contract of employment, which was drawn up in Paris, detailed how
his future work should specifically centre on supporting the bourgeoning
Swedish silk industry. The instruction defined that Rehn should design
all patterns that the Swedish silk manufactures needed to develop, to su-
pervise that the patterns were transferred in a proper manner during the
weaving and to review the end result. A clear model was presented when
Hårleman pointed out Lyon, the preeminent French centre of silk production, as the place where Rehn should go to secure the necessary information.

In a detailed instruction, Rehn was ordered to go to the town, visit its large manufactures, note everything essential, secure all designs, patterns and shapes worthy of imitation, gain knowledge of all the processes in the production and take drawings of the most reputable machines.[17]

By this point a first group of French silk weavers had already arrived in Sweden and set up their manufactures. In the end at least five of the 11 new silk weaving manufactures that started operations in Stockholm between 1739 and 1746 were founded by French silk weavers, who constituted the fully dominant foreign influence on the industry. Out of the group the already mentioned Bartolomei Peyron, who established his manufacture in 1741, remains by far the best known. This not only was due to the fact that he by all accounts was the most accomplished artist, but also because he arguably turned out to be the most successful manager of his enterprise.[18]

Indications of an adaptive industry 1744–1831

Peyron's manufacture was located in a property on Norrmalm, north of the city centre, which in 1752, in another clear indication of the official appreciation of his work, was gifted to him by the Manufacture Office.[19] The plant remained operational under his direction until 1760, at which point it was sold to the brothers Jan, Abraham and Sophonias Alnoor. At the time of the transaction, the manufacture was at an absolute peak, with no less than 110 looms in operation, having expanded from 56 looms, five years before.[20]

Despite its apparent size and growth, Peyron's plant was neither the biggest silk weaving manufacture in Stockholm in 1760 nor the most expansive. This honour befell the industrialist Anders Dahlmansson, who already in 1745 operated 45 looms in a manufacture on Södermalm, south of the city centre. Between 1755 and 1761 the size of his operation almost doubled when the number of looms rose from 108 to 202.[21] Dahlmansson most likely benefitted from the new state support policies, even though it is unknown exactly how his expansion was financed. It is at the same time clear that he started from a well-established position. The business, which he managed since the early 1740s, had been founded by the merchant Johan Efwerling in 1725 and was operated from the same site from 1731 at the latest. It was thus one of the first silk weaving manufactures that established in the eighteenth-century rebirth of the industrial sector.[22]

What happened next, after the peak years around 1760–1761, was an extremely dramatic downturn that affected the entire silk weaving industry. In the old Peyron manufacture, now under the control of the Alnoor brothers, the capacity was cut from 110 looms in 1760 to a mere 46 in 1764. Dahlmansson led the way, however, when slashing the number of looms in his manufacture from 200 to 50 between 1762 and 1763.

The general slump has traditionally been seen as the result of an international financial crisis that hit Sweden hard between 1763 and 1765 and affected commerce and trade, including some of the biggest merchants and wholesalers. The slump was however long term also affected by the new commercial ideas that were implemented following the downfall of the Hat Party at the Diet of 1765–1766. The new political leadership that came to power shifted focus away from supporting domestic manufactures. This included closing down the Manufacture Office as well as removing the system of subsidies.

While the period after 1765–1766 saw the establishment of a new number of enterprises in the silk weaving industry (see Figure 6.1), the total number of looms remained relatively low, never approaching the situation in the early 1760s, when the number of looms was at an absolute peak.

The new, smaller silk weaving manufactures were partly a result of the removal of previous restrictions on industrial establishments. After 1766, not only individuals who had received rights to operate as industrialists but also the actual artisans, the weavers, who had previously been employed by the industrialists were allowed to establish their own operations. As a result, a number of small-scale operations were established during the following decades.

The greater number of enterprises did not mean that the number of looms per enterprise, when viewed as an average, dropped much. What seems to have happened instead was a stabilization of the typical size of the manufactures, with the majority of the enterprises operating between 10 and 30 looms (see Figure 6.2). No really large and few very small enterprises existed during this period.

This seemingly did not mean that the economic behaviour became sounder. The period between 1780 and 1790 on the contrary saw an unusual number of bankruptcies, which overall eclipsed the situation in and around the mid-1760s or for that matter the period between 1805 and 1815 (see Figure 6.3).

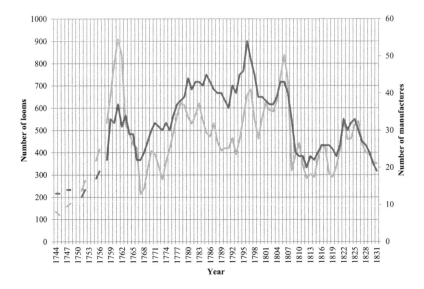

Figure 6.1 The number of silk weaving manufactures and associated looms in Stockholm, 1744–1831

Note: Dark grey: Number of silk weaving manufactures; Light grey: Number of looms.

Source: RA, Kommerskollegium, Kammarkontoret, Da1, Årsberättelser fabriker serie 1, 1744–1831.

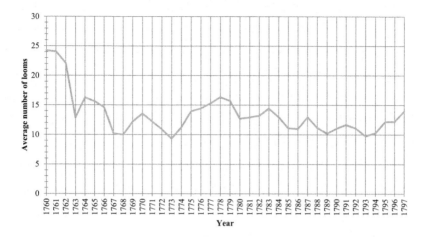

Figure 6.2 The average number of looms per silk weaving manufacture in
Stockholm, 1760–1797

Source: RA, Kommerskollegium, Kammarkontoret, Dal, Årsberättelser fabriker serie 1,
1760–1797.

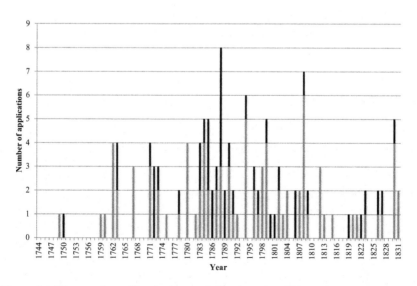

Figure 6.3 The number of bankruptcy applications by silk weaving manufacturers
and silk weaving artisans in Stockholm, 1744–1831

Note: Light grey: the number of applications by silk weaving manufacturers; Dark grey: the
number of applications by silk weaving artisans (silk weavers, apprentice silk weavers,
master silk weavers and silk dyers).

Source: www.tidigmodernakonkurser.se.

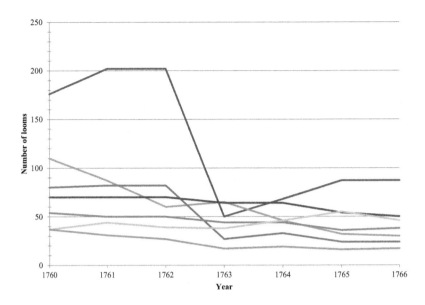

Figure 6.4 The number of looms in the seven largest (30+ looms) silk weaving manufactures in Stockholm, 1760–1766
Source: RA, Kommerskollegium, Kammarkontoret, Da1, Årsberättelser fabriker serie 1,
1760–1766.

In this perspective, the development in the mid-1760s, while dramatic, far
from constituted a downfall on a scale that led to the collapse of the entire
industry and a range of interconnected bankruptcies. What seems to have
happened instead was a general downsizing, or adaptation to a new situation, with a smaller number of bankruptcies among the smallest and most
vulnerable operators. When broken down on an individual level, the data
show not only a surprising similarity in reaction from almost all established
players involved in the silk weaving industry but also a similar return to a
state of normalcy after the crisis passed (see Figure 6.4). Whereas people
like the Alnoor brothers and Dahlmansson dramatically cut the production
capacity, this seems to have been a conscious (and ultimately successful)
move to adapt rather than a desperate step to survive.

I would argue that a similar development occurred during one of the other
major economic shifts with a wide impact on the Swedish economy and the
industrial landscape, that has been identified in previous research, namely
the time-period around 1805–1815. The most substantial crisis during this
period happened in 1808–1809 during the Finnish War, a conflict fought
between Sweden and Russia over the control of Finland. In the Treaty of
Fredrikshamn signed in September 1809 to end the war, Finland was permanently ceded to Russia.

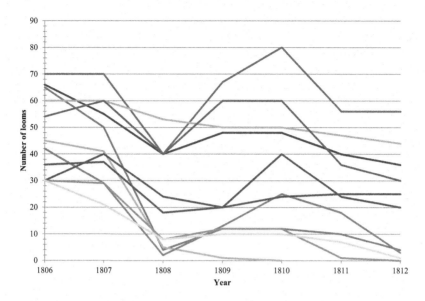

Figure 6.5 The number of looms in the 11 largest (30+ looms) silk weaving manufactures in Stockholm, 1806–1812

Source: RA, Kommerskollegium, Kammarkontoret, Dal, Årsberättelser fabriker serie 1,
1806–1812.

The effects on Swedish society, politics and the economy were substantial.
Most notably, it led to the overthrow of King Gustav IV Adolf (r. 1792–1809)
and the ascension of his uncle King Karl XIII (r. 1809–1814) to the throne.
Sweden furthermore was forced to switch sides in the larger Napoleonic
conflict. It joined the Continental System, the blockade policy imposed by
France and Russia against the United Kingdom. The results were wide-
ranging, limiting access and trade links to England, but foremost cutting
off the Finnish market to merchants and industrialists. The loss of Finnish
contacts and the near impossibility to get back money owed by individuals
in the area can also be seen in the argumentation by some of the industrialists who declared bankrupt in the wake of the crisis. The majority of the established silk weaving manufacturers never declared bankrupt or went out
of business, however. Much like during the crisis in the 1760s they instead
managed to react and adapt to the impact of the war (see Figure 6.5).

Cooperation and resilience

All in all I would argue that the silk weaving industry during both periods
of crisis showed a high degree of adaptability and resilience, which helped
it through the development. One of the key reasons was the close-knit,

integrated character of the industry. This included a number of factors. The silk weaving and silk processing industry for one was largely confined to a few relatively confined geographical locations in Stockholm. This included two areas on the island of Södermalm, south of the city centre and a similar area on Norrmalm, north of the city. The concentration facilitated communication and the exchange of information and ideas, the forging of contacts within the industry and the development of social connections. The clustering of businesses, in this respect, did not create negative competition or the inability of individual enterprises to survive during periods of turmoil, but on the contrary supported adaptation and a resource pool with a high degree of interconnectedness and exchange.

Silk weaving and silk processing enterprises showed a very high degree of connectivity during the period, including mergers and the transfer and sale of operational plants. The sector also saw a lot of intermarriage between families with a long history in the business. It widely accepted and seemingly supported widows taking over the operation from their deceased husbands, again supporting the continuous operation of the manufactures, rather than accepting their demise.

The silk weavers or, in other words, the actual artisans who worked in the manufactures shared a collective identity and were organized in a silk weavers' association, which differentiated them from other individuals who worked with silk on a smaller or different scale, including silk stocking weavers and silk ribbon makers. The cooperation and organization within the industry and the geographical concentration likely meant that the artisans were well-known, or otherwise easy to research, which helped when recruiting, by simplifying the judgement of skills and personal traits.[23]

The silk weaving manufacturers, as employers and financiers, at the same time were highly integrated horizontally by working together or in close cooperation with various other groups in society. This included other industrialists who provided the silk yarn that was needed for the weaving, patternmakers, who developed the various weaves that were incorporated in the textiles as well as agents and other salespeople who handled the sale of the finished textiles. Close contacts meant a good knowledge of the market, flexibility and ability to quickly adapt.[24]

The shifting nature of a silk weaving manufacture

The first years

How this flexibility played out in practice will be traced in the following through an investigation of the rise and fall of a silk weaving manufacture originally founded in 1758 by Hans Magnus Furubom, but, as will be seen, with a life-span stretching far beyond his time, well into the 1830s. In the following, I will build on the argument about adaptability and resilience by showing how the manufacture survived despite the various crises that affected the economy, including the often-discussed general stagnation which affected Stockholm during the period.

The Furubom manufacture originally started production and delivered textiles for inspection to the municipal control authorities in 1758. It was only formalized in 1760, however, after Furubom gained burgher rights as a silk manufacturer. At the same time, he entered into a partnership with Per Carlsson Berger, after which the enterprise was referred to as Furubom & Berger.[25]

The manufacture was originally established in a city parcel on western Södermalm. The rented property in which the plant was located was shared with a number of other businesses. In 1760 these included two other silk weaving manufactures, founded in 1755 and in 1759, respectively. A fourth silk weaving plant was located in an adjacent property in the same city parcel. The concentration was no coincidence. The area, as mentioned, constituted one of three relatively confined geographical locations in the Stockholm where the majority of silk weaving operations were operating during this period. In 1760, at least 12 manufactures related to silk weaving or silk processing were situated close to Furubom & Berger's manufacture.[26]

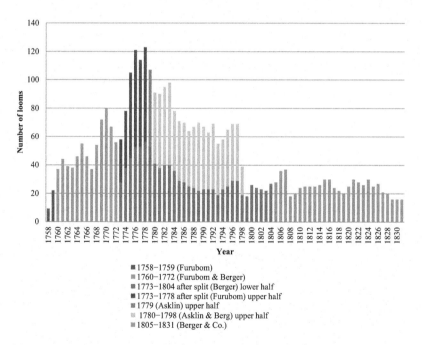

■ 1758–1759 (Furubom)
■ 1760–1772 (Furubom & Berger)
■ 1773–1804 after split (Berger) lower half
■ 1773–1778 after split (Furubom) upper half
■ 1779 (Asklin) upper half
▨ 1780–1798 (Asklin & Berg) upper half
■ 1805–1831 (Berger & Co.)

Figure 6.6 The shifting ownership and the number of looms in a silk weaving manufacture founded by Hans Magnus Furubom in 1758

Source: RA, Kommerskollegium, Kammarkontoret, Dal, Årsberättelser fabriker serie 1, 1758–1831.

Building on earlier roots

The manufacture grew quickly from 37 looms in 1760 to an absolute peak with 80 operational looms a decade later (see Figure 6.6). Although affected by the unstable conditions during the financial crisis of 1763–1765, the manufacture carried on largely unscathed and continued to expand. As part of the growth, by the end of the decade, the operation was moved to a city parcel on eastern Södermalm, in a property that was acquired from the wholesale merchant Georg Friedrich Diedricksson.[27]

The relocation was strategic. The new property was one in a number of properties that had been formerly owned by the silk manufacturer Anders Dahlmansson. As mentioned earlier Dahlmansson had operated a silk weaving plant at the site from the early 1740s until 1767.

The acquisition not only allowed Furubom & Berger to build on the heritage of one of the oldest silk weaving operations. It also placed their manufacture in the middle of a uniquely dynamic area. This was basically caused by the successful operation of the earlier plant, which under Dahlmansson, was expanded into the biggest ever silk weaving enterprise in Sweden. At its peak in the years 1761–1762 when it operated 202 looms, it had a workforce of 582 workers, equalling more than a quarter of the total workforce in the entire business sector. The surrounding geographical area in the process became one of the most important centres of silk production in the capital. The area included a number of other plants, places of retail and the highest concentration of industrial masters, experts and workers with competence in silk weaving and processing. As mentioned, Dahlmansson managed the international economic crisis of 1763–1765 reasonably well, after radically slashing capacity from 202 to 50 looms between 1762 and 1763, while at the same time applying for bankruptcy protection.[28] His radical approach and ability to reach an agreement with his creditors assured the survival of his operation, which in the period 1763–1766 was expanded to 87 looms, thereby also retaining its position as the most significant silk weaving operation in the capital.

In 1767, the operation came to a halt, however, after Dahlmansson's sudden death, which forced his wife and heirs to file for bankruptcy to manage the outstanding debts.[29] One of the results of this process was the sale of the three buildings in which the manufacture was located, at a public municipal auction. The factory plot that was bought by Diedricksson and then resold to Furubom & Berger has already been mentioned. The two other transactions transferred a second factory plot and Dahlmansson's main residence. Both properties were acquired by Henrik Hackson, a former Swedish consul in Smyrna, but were immediately handed over to the silk ribbon manufacturer Rudolf Stenberg.[30]

The strategy behind the Hackson–Stenberg transaction again reveals a great deal about the close-knit character of the Stockholm silk industry both in terms of family connections and investments. Stenberg had no background as a silk ribbon manufacturer. When he acquired burgher rights in Stockholm in 1761, he did so as a textile retailer.[31] His marriage the year before had however enabled him to establish a close relation with local silk ribbon manufacturers. His wife was the daughter of a Dutch silk ribbon manufacturer, who had arrived in Stockholm a decade earlier. From 1755 until 1765 his wife's brother operated the family silk ribbon manufacture, which by 1760 was located in a city parcel east of Stenberg's future manufacture. In 1755, another of Stenberg's brothers-in-law had also established a silk ribbon manufacture which in 1758 was operating from a property right next to the Dahlmansson manufacture in the same city parcel.[32]

When this second brother-in-law died in 1760, Stenberg took over the operation of his manufacture.[33] As a result, when Hackson transferred the ownership of the two former Dahlmansson properties to Stenberg in 1768, he in effect added real estate to an already existing and well-established manufacture in the adjacent plot in the same city parcel.[34] The new properties were apparently considered superior, as the old property was put up for sale and disposed of by Stenberg the following year. On a general level, the benefits of the transactions were clear. By taking over properties in immediate proximity to an existing operation, it was possible to make the most of established networks as well as gain access to workers and supporting industries in the surrounding area.

It is also likely that the process was founded in a financial investment. When reviewing Dahlmansson's bankruptcy proceedings, it can be determined that both Diedricksson and Hackson were two of Dahlmansson's major creditors and hence likely to have taken over the three properties as payment to cover unpaid debts.[35] In the case of Hackson, this had a further dimension. Hackson in fact had an earlier history of investments in the silk industry.

In 1759, he both invested in and accepted the role as sales agent for a silk weaving manufacture founded the previous year by the French-born silk weaver Jean Baptist Coudurier. Further underscoring the geographical clustering of the industry, this plant was located in the same area on western Södermalm where Furubom in 1758 had established his first plant. When Coudurier filed for bankruptcy in 1760, Hackson took over the enterprise, but immediately wound it down to be able to reclaim unpaid debts.[36]

It is not fully clear if Hackson's involvement in Dahlmansson's enterprise was as an investor or as a combined investor and sales agent. In the subsequent case involving Stenberg, it becomes clear, however, that his involvement went well beyond acting as a sales agent. This can be understood not least from the substantial values which changed hands immediately following the auction in 1768. Whereas the factory plot was acquired for a relatively modest 6,000 dollars copper money, the main residence was acquired for 60,000 dollars copper money, a substantial sum which suggests that Hackson rather than selling the properties, instead handed them over to Stenberg, as part of an investment in the silk weaving enterprise.[37]

Longevity of operations

Taken as a combined whole, the Stenberg and Furubom & Berger establishments in the same city parcel transferred the assets and properties of Dahlmansson's silk weaving enterprise into new hands. The development ensured continued operations, but also effectively cemented older patterns of cooperation and geographical clustering, while at the same time creating new opportunities. One of the key aspects of this cooperation was the fact that the manufacturers could operate, not as competitors, but rather in a complementary way.

Dahlmansson's operation had included a silk weaving manufacture, but also a silk processing, or more precisely a silk throwing plant. The latter operation, through which raw silk was refined into silk that could be utilized for weaving was not an exclusive in-house operation. Dahlmansson, as well as other silk throwers, also refined silk that was used by other plants. The elimination of his silk throwing plant in 1767 thus was less than ideal for the surrounding enterprises that lacked the capacity to refine silk. Furubom & Berger quickly moved to restart the particular operation and already in 1769 produced processed silk yarn at the silk throwing plant again.[38]

Both the Stenberg and Furubom & Berger manufactures thrived at the new location. Stenberg managed his manufacture until his death in 1803. By this point, the practical operations had already been taken over by his son Emanuel. After formally buying the properties from his mother and other heirs after his father's death, Emanuel continued to operate the manufacture until his own death in 1834, managing the operation throughout the crisis of 1805–1815. At the time of his death, a continuous production of silk ribbons had been conducted at the site for almost 80 years.[39]

The Furubom & Berger enterprise also showed a marked longevity. Unlike the Stenberg manufacture which continued as a family enterprise, the manufacture operated by Furubom & Berger changed hands on a number of occasions. A significant change happened already in 1773 when the two partners split the enterprise into two entities, with Furubom taking over the operation of 30 and Berger 28 of the existing looms. The decision was amiable. This was further underscored by the fact that the two partners at this point were interlinked not only professionally, but also through family, after Furubom had married Berger's sister.[40]

Berger continued to operate his part of the silk weaving enterprise from 1773 until his death in 1802, in later years in cooperation with his two sons. After the split, he relocated the manufacture to a city parcel on western Södermalm. In 1781, an adjacent property at the new site was bought by the silk weaving manufacturer Carl Jacob Ström. The purchase re-established old contact surfaces. Ström was in fact one of the other silk weaving manufacturers who twenty year before had worked in the same property as the, at that point, newly established Furubom & Berger manufacture.[41]

A renewed cooperation between the old neighbours can be hinted at when considering how Berger's sons in 1804, took over a different property, also on western Södermalm, where Ström had operated his silk weaving manufacture from 1779 until his death in 1800.[42] The acquisition documents however show that the purchase agreement for the property had been signed already in 1797, when both Ström and Berger were still alive, in a way which suggests a planned transfer of operation of the silk weaving enterprise. The Berger silk weaving manufacture relocated to the site of the old Ström manufacture in 1804, where it continued its operation under the direction of Berger's sons, and subsequently, their heirs until 1838, without any indications of problems arising from the crisis of 1805–1815.[43]

Incorporating a French heritage

Furubom remained on the old Dahlmansson-property, after buying out Berger in 1774.[44] In 1775, he added substantially to the operation when he took over the ownership of a silk weaving operation operated by Johan Anders Meurman and merged his 20 looms into the existing operation. The move yet again shows the continuity, integration and cooperation within the industry. The plant which he acquired was situated in a property in a city parcel, literally around the corner from his existing factory.

The new manufacture had a notable history that stretched back to 1746, or in other words to the same decade during which Dahlmansson had started his operation. What Furubom acquired was a manufacture that had been originally established by Jean Poyet, one of the five French silk weavers who established operations in Stockholm between 1739 and 1746. By 1755, at the latest, Poyet operated his manufacture from the property around the corner from Dahlmansson. The following year, the operation was acquired by the merchants Sten & Thuring (see Figure 6.7).[45]

Between 1760 and 1765, the production at the factory was registered on the new business partners, before the operation again shifted hands in 1766 and the production was taken over by Meurman. Neither Sten & Thuring nor Meurman appears to have managed the financial crisis of the 1760s very well, and the operation experienced a steady decline during this period from 54 looms in 1760 to 14 looms in 1773. An indication of the situation can be had from an application for bankruptcy which was handed in by Meurman in 1771, but subsequently recalled, suggesting a settlement out of court.[46] Whether Furubom was involved in this is unclear. His subsequent acquisition of the enterprise and its incorporation into his existing operation still at any rate ensured the survival of one of the earliest and arguably more illustrious silk weaving manufactures in Stockholm, albeit under new leadership.

Furubom would leave the business a few years later. In 1779 he handed over the enterprise including the silk weaving manufacture and the silk throwing plant to Jonas Asklin, who the following year paired up with Anders Berg to run the enterprise as Asklin & Berg. The new owners bought the property from Furubom in 1781.[47]

The silk weaving enterprise remained operational at the site until 1799. At this point, it was dissolved due to what was described as "intervening

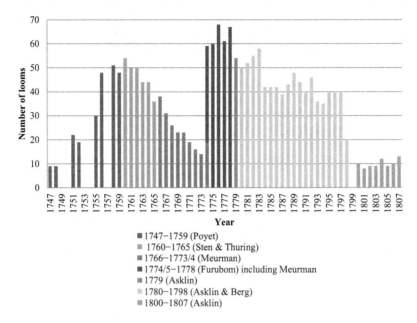

Figure 6.7 The shifting ownership and the number of looms in a silk weaving manufacture founded by Jean Poyet in 1746

Source: RA, Kommerskollegium, Kammarkontoret, Dal, Årsberättelser fabriker serie 1,
1747–1807.

obstacles".[48] What this meant, and the extent of the operation at the time
was expounded on in a voluminous bankruptcy application delivered to the
Stockholm Town Hall Court soon thereafter.[49]

The property in which the manufacture was located was denoted as in
possession of the bankruptcy administrators in 1800, but was subsequently
handed back to the manufacturers. Asklin later that same year took full possession after an agreement that saw Berg hand over his rights to the deed to
his business partner. This also appears to have paved the way for a relaunch
of a much-reduced silk weaving manufacture at the site under the sole ownership of Asklin. The new enterprise remained operational until 1806. At
this point, it was handed over to Carl Fredrik Mancke, who in late 1805 had
entered into an agreement with Asklin to buy the factory plot.[50] Although
the production continued, it came to an abrupt halt after Mancke travelled
to Hamburg without returning. This was followed in 1807 by the death of his
wife. The turn of events meant that silk weaving operations at the property
also finally ended after more than 65 years of continuous operation.[51]

Adjusting capacity: the role of the looms

Next to the networks, geographical clustering and various economic and social
strategies outlined above, I would argue that the silk weaving manufacturers

in Stockholm also proved resilient because they were able to adapt to sudden crises by quickly adjusting their capacity by activating or deactivating looms.

The silk weaving industry utilized a number of different technologies to turn the silk yarn that was bought from yarn makers or directly from foreign suppliers into textiles. The silk throwing process, including reeling, throwing and doubling, during which processed raw silk was further treated and arranged in order to achieve the needed strength for weaving was undertaken partly at the yarn makers, but at the bigger silk weaving plants more typically in house, both for internal use and for sale to smaller manufactures. The by far biggest operation at the actual silk weaving plants, however, and next to the raw-material the biggest investment for a manufacturer, was the actual weaving and the looms on which the weaving was undertaken. The looms not only were the most complicated pieces of machinery in the bigger processing and manufacturing of silk textiles, but also the most expensive fixtures and also required the largest floor space. Before the introduction of the Jacquard loom (which only entered the Swedish silk weaving industry around 1830), they also required a substantial workforce to operate.[52]

Importantly, this workforce was to a large degree made up of fairly unqualified helpers, whereas only a set number of operators at the weaving manufactures were artisans. While the latter individuals, including master and apprentice weavers, were highly valued and also supported by the industrialists in financially difficult times, the bulk of the workforce, who performed the more menial or repetitive tasks, seems to have been considered much more replaceable. This also meant that the costs of the workforce rarely weighed down the operation of the enterprise to any large degree. When structural crises or other problems arose, or when normal market demand was slow, the unqualified helpers, as well as weavers working off-site on commission could quite simply be laid off with little long-term impact on the knowledge base in the operation and limited economic fallout.

In the short term, looms could also be deactivated. Technically, this was not as easy. A characteristic aspect of a loom in operation was the fact that much of the actual raw material was fixed to the frame. This for one meant that a loom in the middle of an unfinished project was highly valuable, but at the same time highly expensive. It also meant that the decision on what to produce was fixed, after the loom was set up and prepared for weaving. Before the loom could be made available for a new project, the ongoing weaving had in short to be fully finished. What this meant for the manufacturers was that the need to finish uncompleted textiles was paramount both to be able to sell the finished product, to be able to get payment for the textile, but also to free up the loom, to be able to start another project. This also meant that it was difficult to take on new orders or start new projects before a loom was ready for operation. Even a more valuable order could not easily be accommodated before an earlier order was finished. The only way to get around this was to invest in new looms.

This structural problem can serve to explain the rapid expansion both of some of the largest manufactures in the period before 1760, but also the smaller, and more variable adjustments in capacity during the following decades. I would argue that one of the characteristics of the silk weaving industry as a whole was the constantly fluctuating number of looms that were operated by individual manufacturers. While the average number of looms across the industry during certain periods remained relatively constant, for example as shown from the late 1760s until 1797 (see Figure 6.2), on an individual level, fluctuations were still apparent. While examples of rapid rises and drops can be found, the development was overwhelmingly not volatile, or concerted, like in times of outside crisis, such as in the periods 1760–1766 or 1806–1812 (see Figures 6.4 and 6.5). It instead took the form of a constant individual fine tuning and adaptation of capacity, likely as a response to shifting market demands.

A problematic aspect of this behaviour was that it basically required excess capacity, which could be utilized when times were good, but was deactivated when the situation so necessitated. That this indeed was the case can be clearly seen when reviewing manufacturers who filed for bankruptcy during the period. Three examples will suffice at this point.

Carl Antonsson operated a silk weaving manufacture from 1811 until declaring bankruptcy in 1815. In his report to the municipal authorities he reported 10 operational looms in the first year of operation. The following year 13 looms were used. According to the same records, during the final two years nine looms were used. The information that was supplied by Antonsson during his bankruptcy proceedings in 1815 however shows a significantly more extensive operation. The detailed inventory of his plant shows that it included no less than 48 looms, 20 looms for taffeta, 12 looms for satin, eight looms for velvet, six looms for the production of shawls and two looms for the production of levantine.[53]

Hans Widte, who declared bankruptcy in 1798, had reported two operational looms to the municipal authorities the previous year. The difficult situation in which he found himself can be hinted at from the fact that nothing was produced on the two looms. This was however much more acute than the numbers suggest. The inventory of his manufacture which was supplied to the Stockholm Town Hall Court during his bankruptcy proceedings showed that he in reality owned 14 looms, 13 ordinary ones and 1 loom that was used for the production of shawls. While Widte had operated eight to nine looms in the early 1790, one has to go back to the mid-1780s to find instances when 14 or more looms had been used in the plant.[54]

Magnus Lindström filed for bankruptcy in 1808. Overall Lindström was the largest player to be forced out of business as a result of the collapse of the market after the outbreak of war with Russia in 1808. During the final year of operation, he had only used two looms. At no point between 1800 and 1807 did he operate fewer than 40 looms however, peaking in 1801 when 48 looms were in use. The inventory of his manufacture that was delivered to the court during the bankruptcy proceedings showed a detailed list of no fewer than 52 looms of various configurations and uses in his possession.[55]

The three examples show quite clearly the substantial overcapacity in the industry. As suggested, this was not necessarily a problem, but rather a quality that benefitted the industry, by creating an inherent flexibility and capacity to quickly make use of market opportunities. Still, when the manufacturers failed in their adjustments and were forced into bankruptcy they struggled with substantial debts that could be traced back to underutilized looms, suggesting a precarious balance between investing in new capacity and managing to cut down on capacity in times of crisis.

Conclusions

This chapter has shown how the silk weaving industry in Stockholm grew from something almost non-existent to a fairly substantial industry in a relatively short span from the 1740s until the 1760s. The development was helped by extensive state support, benefitted from royal patronage and utilized French expertise and ideas. In the long run, after the removal of state subsidies, the industry settled into a different type of environment, which required new approaches for survival.

I have underlined how this led to strategies which included extensive use of professional and social networks, benefits drawn from geographical concentration, horizontal and vertical integration and cooperation, which combined served to secure long-term survival. The investigation suggests that the industry during the period was surprisingly adaptive to suddenly changing market conditions in the form of wars and economic crises that affected demand or supply. As a result, it was also surprisingly resilient, at least in a short-term perspective. The entire industry from the biggest to the smallest enterprises in times of crisis seems to have reacted in similar ways, when reducing capacity by slashing the number of operational looms.

The reason for the seemingly coordinated response most likely was the financial dependence between the various enterprises, the fact that the entire industry produced for the same local market and likely had a shared identity and similar viewpoints. By swiftly identifying problems and collectively reducing capacity, production could be halted and unwanted supply avoided. The result, it seems, was a long-term survival of the majority of the players in the industrial sector. When problems arose, the close coordination meant that substantial efforts were made to find solutions that would help ailing enterprises to survive. This included mergers and the temporary administration of the enterprises by creditors with the intent to continue production.

What also marked the industry was a constant tendency to adjust the capacity by adding or reducing the number of operational looms during relatively normal market conditions, when no apparent large-scale crises affected the situation. The observation becomes particularly interesting when viewed in relation to the observations about a new type of production and ultimately a new adaptation to a broader market segment, in the period after the 1760s. The rise of a new type of production, more adapted to the new luxury demands of broader consumer groups, was seemingly accompanied by an equally adaptable industrial structure, with manufacturers who were embedded in a complex structure of social networks that enabled them to quickly adapt and fine tune their production to fit new market conditions.

Notes

1 RA, Kommerskollegium, Huvudarkivet, DVIIIa, Förteckningar över utfärdade fabriksprivilegier, vol. 1. For information about another silk weaving manufacture that was established in Stockholm towards the end of the seventeenth century, see Berg and Stavenow-Hidemark (2001).
2 The statistical information about the number of enterprises presented here and in the following has been compiled from the voluminous unpublished industrial accounts from the period, with separate volumes covering most years from 1744 onwards. RA, Kommerskollegium, Kammarkontoret, Da1, Årsberättelser fabriker serie 1, 1744–1831.
3 For examples of studies of the silk textiles produced in the Swedish manufactures, see Wallin (1920); Sylwan and Geijer (1931); Berg and Stavenow-Hidemark (2001); Ciszuk (2012); Stavenow-Hidemark and Nyberg (2015).
4 Heckscher (1937), pp. 153–221; Heckscher (1935–1949), part II, vols. 1–3.
5 Nyström (1955); Kjellberg (1943); Nyberg (1992, 1999b); Persson (1993).
6 Nyström (1955), pp. 172–173.
7 For an English language treatment of the Age of Liberty, see Roberts (1986). Various aspects of the political culture during the period are covered in Skuncke and Tandefelt (2003).
8 The biography of the Queen in Jägerskiöld (1945). Her efforts among other things led to the arrival of a French theatre company in 1753 and subsequently to the creation of a permanent court theatre (Beijer 1981).
9 The outfit ultimately worn by the King was woven with a pattern of open crowns and silver flames on a silver and white silk background, whereas the elaborate French court dress worn by the Queen was woven from similarly coloured brocade silk, decorated with golden crowns and silver stars. For the Queen's dress placed in the context of similar robes, see Rasmussen (2019a, 2019b). For Peyron's role in the production, see Stavenow-Hidemark and Nyberg (2015), pp. 224–225.
10 Hildebrand (1874–1875), vol. 2, pp. 119–122. When Peyron died in 1766, the gold medal was listed in his probate inventory. SSA, Rådhusrättens 1:a avdelning (Justitiekollegium, Förmyndarkammaren), F1A, Bouppteckningar, vol. 207 (1766:3), pp. 830–831.
11 Generally, for the Queen and her role as a collector and patron of the arts, see Laine (1998). For details on her silk experiments, see Barton (1977), pp. 81–98; Barton (1985, 1994). On this and other attempts to establish silk production in Sweden during the eighteenth century, see Johansson Åbonde (2010), pp. 23–58. A brief overview of these attempts can be found in Marsh (2020), pp. 221–224.
12 A precedence for using French artisans at Stockholm Castle had been established when a group worked at the site between 1693 and 1713 (Hinners, 2012).
13 Generally, for the building works at Stockholm Castle during this period, see Olsson and Böttiger (1940–1941), vol. 3. The role of Louis Masreliez and Jacques-Philippe Bouchardon, two of the most prominent artisans involved in the project, is covered extensively by Moselius (1923) and Lindblom (1924), respectively. For the role played by Hårleman, see Stavenow (1927); Alm (1993).

Economic behaviour and social strategies 117

14 Generally about Tessin, see von Proschwitz (1995), and for his political role, see Nilzén (2012). For his activities during his years in Paris, see Leijonhufvud (1915); von Proschwitz (1983, 2002). For his art collection, see Heidner (1995), pp. 21–40, and generally for the French influence on the Swedish cultural elite during the Age of Liberty, see Wolff (2005).
15 Ilmakkunnas (2015, 2017).
16 About Rehn and his role in the 1750s, see Olsson (1932, 1945). For an overview of the research about his life and artistic role, see Engellau-Gullander (2001).
17 Malmborg (1927), pp. 108–109. Generally for the type of industrial espionage in Lyon which Rehn was instructed to engage in, see Bertucci (2013), pp. 820–852.
18 Two letters of his hand, written in Stockholm, in: RA, Ericsbergsarkivet, Auto-grafsamlingen, vol. 162, fr. Peyron 1741-10-10, 1742-04-08.
19 SSA, Stockholms magistrat och rådhusrätt, A6a, Lagfarts- uppbudsprotokoll, vol. 33 (1750–1752), 1752, pp. 212, 230, 249.
20 Abraham Alnoor had operated a small silk weaving manufacture since 1755, whereas Sophonias had been employed in the Peyron manufacture. After tak-ing over the new manufacture, the brothers continued to operate it at the site until 1790, albeit gradually at a much reduced size. SSA, Överståthållarämbe-tet för uppbördsärenden, Kamrerarexpeditionen, G1BA, Mantalslängder, 1760 (Jakob), p. 107; 1770 (Jakob), p. 128; 1780 (Jakob), p. 134; 1790 (Jakob), p. 222; Stavenow-Hidemark and Nyberg (2015), pp. 235–236.
21 Examples of the silk that was produced at the manufacture can be seen in Stavenow-Hidemark and Nyberg (2015), pp. 189–193.
22 Efwerling acquired burgher rights as a merchant on 1725-09-09. SSA, Handel-skollegiet, D1a1, Borgarböcker, huvudserie, vol. 1 (1689–1750), p. 167. He would eventually become the owner of two properties in the city parcel *Kejsaren* that he purchased in 1731 and 1739, respectively. SSA, Stockholms magistrat och rådhusrätt, A6a, Lagfarts- uppbudsprotokoll, vol. 26 (1731), 1731, pp. 104, 110, 119; vol. 29 (1738–1740), 1739, pp. 207, 238, 269. In 1740, he was still listed as the owner of the manufacture. RA, Kommerskollegium, Kammarkontoret, Da1, Årsberättelser fabriker serie 1, 1740.
23 The detailed knowledge of the different artisans and workers in the textile manu-factures is apparent in the population registry (*mantalslängder*) from the period, which lists not only the profession of the artisans, and their residence, but often also the manufacturer by whom they were employed. SSA, Överståthållarämbetet för uppbördsärenden, Kamrerarexpeditionen, G1BA, Mantalslängder.
24 For the concept of flexibility in the context of Western industries and industri-alization, see the various contributions in Sabel and Zeitlin (1997), including in particular Carlo Poni's study of the silk industry in Lyon (Poni, 1997, pp. 37–74).
25 Both acquired burgher rights to work as silk manufacturers on 1760-07-14. SSA, Handelskollegiet, D1a1, Borgarböcker, huvudserie, vol. 2 (1750–1829), pp. 19, 94.
26 The property (no. 115) in which the plant was situated, together with the adja-cent properties (nos. 116 and 117), in the city parcel *Rosendal*, were all owned by the merchant ship captain Magnus Ahlström. SSA, Överståthållarämbetet för uppbördsärenden, Kamrerarexpeditionen, G1BA, Mantalslängder, 1760 (Ma-ria Magdalena inre), pp. 56–57, 63. The two other silk weaving manufactures in the property were operated by Carl Jacob & Olof Bernhard Ström and Lars Sil-jenstedt & Pehr Ruth, respectively. The plant in the adjacent property (no. 117) was operated by Erland Noreén, who did this without an official concession. Sil-jenstedt, much like Furubom and Carlsson Berger had acquired burgher rights as silk manufacturers on 1760-07-14 whereas Per Ruth only did so on 1764-01-09. SSA, Handelskollegiet, D1a1, Borgarböcker, huvudserie, vol. 2 (1750–1829), pp. 264, 283.

118   *Håkan Jakobsson*

27 The property in the city parcel *Kejsaren* was acquired for 10,000 dollars copper money in August 1768. SSA, Stockholms stadsingenjörskontor, Förrättningsprotokoll, Designationer och planritningar, A1c, Designationer, vol. 10 (1767–1773), pp. 11, 116.

28 The application was made on 1763-06-04. SSA, Stockholms magistrat och rådhusrätt, C5a, Konkursdiarier, vol. 1, p. 17.

29 Two applications were made on 1767-08-10 and on 1767-11-25. SSA, Stockholms magistrat och rådhusrätt, C5a, Konkursdiarier, vol. 2, p. 15; vol. 3, p. 4. A verdict was delivered on 1769-03-11.

30 In December 1767, i.e. before the auction, property no. 167 had already been handed over to the creditors. SSA, Stockholms stadsingenjörskontor, Förrättningsprotokoll, Designationer och planritningar, A1e, Designationer över ofria tomter, vol. 14 (1765–1770), pp. 349, 357; A1c, Designationer, vol. 10 (1767–1773), p. 10.

31 SSA, Handelskollegiet, DIa1, Borgarböcker, huvudserie, vol. 2 (1750–1829), p. 284.

32 SSA, Överståthållarämbetet för uppbördsärenden, Kamrerarexpeditionen, G1BA, Mantalslängder, 1760 (Katarina norra), pp. 157, 212; SSA, Stockholms magistrat och rådhusrätt, A6a, Lagfarts- uppbudsprotokoll, vol. 35 (1756–1758), 1757, pp. 155, 170, 190; RA, Kommerskollegium, Kammarkontoret, Da1, Årsberättelser fabriker serie 1, 1760, p. 37.

33 RA, Kommerskollegium, Kammarkontoret, Da1, Årsberättelser fabriker serie 1, 1760, p. 37; 1762, p. 73.

34 SSA, Stockholms stadsingenjörskontor, A1c:10 (1767–1773), p. 110; SSA, Stockholms magistrat och rådhusrätt, A6a, Lagfarts- uppbudsprotokoll, vol. 39 (1768–1770), 1769, pp. 68, 81, 104.

35 SSA, Stockholms magistrat och rådhusrätt, C5a, Konkursdiarier, vol. 2, p. 15; vol. 3, p. 4.

36 SSA, Stockholms magistrat och rådhusrätt, F6a, Konkursakter, huvudserie, 229/1780.

37 SSA, Stockholms stadsingenjörskontor, Designationer och planritningar, A1e, Designationer över ofria tomter vol. 14 (1765–1770), p. 357; A1c, Designationer, vol. 10 (1767–1773), p. 10; SSA, Stockholms magistrat och rådhusrätt, A6a, Lagfarts- uppbudsprotokoll, vol. 39 (1768–1770), 1768: pp. 57, 86, 125.

38 RA, Kommerskollegium, Kammarkontoret, Da1, Årsberättelser fabriker serie 1, 1769, pp. 65–66.

39 SSA, Rådhusrättens 1:a avdelning (Justitiekollegium, Förmyndarkammaren), F1A, Bouppteckningar, vol. 353 (1803:3), p. 383; vol. 476 (1834:2), p. 968; SSA, Överståthållarämbetet för uppbördsärenden, Kamrerarexpeditionen, G1BA, Mantalslängder, 1770 (Katarina norra), pp. 173, 179; 1780 (Katarina norra), pp. 200, 206; 1790 (Katarina norra), pp. 232, 238; 1800 (Katarina norra), pp. 233, 242; 1810 (Katarina norra), p. 178; 1820 (Katarina norra), pp. 346, 356; 1830 (Katarina norra), pp. 338, 349. For the transaction that transferred the plots to Emanuel Stenberg in 1803, see SSA, Stockholms magistrat och rådhusrätt, A6a, Lagfarts- uppbudsprotokoll, vol. 50 (1801–1803), 1803, pp. 282, 327, 370.

40 SSA, Rådhusrättens 1:a avdelning (Justitiekollegium, Förmyndarkammaren), F1A, Bouppteckningar, vol. 224 (1772:2), pp. 29–30.

41 Berger's earliest ownership of the property can be established from two architectural drawings. SSA, Byggnadsnämndens expedition och stadsarkitektkontor, F1aa, Bygglovsritningar 1713–1874, 1771:11, 1771:11a. For Ström's purchase, see SSA, Stockholms magistrat och rådhusrätt, A6a, Lagfarts- uppbudsprotokoll, vol. 43 (1780–1782), 1781, pp. 92, 126, 154, and for information about his continued residence at the site, see SSA, Överståthållarämbetet för uppbördsärenden,

Kamrerarexpeditionen, G1BA, Mantalslängder, 1790 (Maria västra), p. 55; 1800 (Maria västra), p. 41.

42 SSA, Stockholms magistrat och rådhusrätt, A6a, Lagfarts- uppbudsprotokoll, vol. 43 (1780–1782), 1781, pp. 48, 66, 104; SSA, Stockholms stadsingenjörskontor, Designationer och planritningar, A1c, Designationer, vol. 16 (1800–1806), p. 416.

43 SSA, Stockholms magistrat och rådhusrätt, A6a, Lagfarts- uppbudsprotokoll, vol. 51 (1804–1806), 1804, p. 536; 1805, pp. 17, 44.

44 SSA, Stockholms magistrat och rådhusrätt, A6a, Lagfarts- uppbudsprotokoll, vol. 41 (1774–1776), 1774, pp. 106, 130, 162.

45 SSA, Stockholms stadsingenjörskontor, Designationer och planritningar, A1c, Designationer, vol. 7 (1756–1759), p. 199; vol. 8 (1759–1762), p. 195.

46 Poyet, Sten & Thuring and Meurman were registered at the address in 1755, 1760 and 1770 respectively. SSA, Överståthållarämbetet för uppbördsärenden, Kamrerarexpeditionen, G1BA, Mantalslängder, 1755 (Katarina södra), p. 389; 1760 (Katarina södra), p. 546; 1770 (Katarina södra), p. 686. The bankruptcy application in SSA, Stockholms magistrat och rådhusrätt, C5a, Konkursdiarier, vol. 2, p. 63.

47 The transfer of the weaving and throwing operations from Furubom to Asklin noted in RA, Kommerskollegium, Kammarkontoret, Da1, Årsberättelser fabriker serie 1, 1779, pp. 6, 53–54. For the acquisition of the property, see SSA, Stockholms magistrat och rådhusrätt, A6a, Lagfarts- uppbudsprotokoll, vol. 43 (1780–1782), 1782, pp. 81, 103, 126.

48 The description in RA, Kommerskollegium, Kammarkontoret, Da1, Årsberättelser fabriker serie 1, 1799, p. 1.

49 SSA, Stockholms magistrat och rådhusrätt, F6a, Konkursakter, huvudserie, 80/1805.

50 SSA, Överståthållarämbetet för uppbördsärenden, Kamrerarexpeditionen, G1BA, Mantalslängder, 1800 (Katarina norra), p. 233. The transfer of the property to Mancke in SSA, Stockholms magistrat och rådhusrätt, A6a, Lagfarts- uppbudsprotokoll, vol. 51 (1804–1806), 1806, pp. 216, 258, 293.

51 SSA, Rådhusrättens 1:a avdelning (Justitiekollegium, Förmyndarkammaren), F1A, Bouppteckningar, vol. 371 (1808:1), pp. 1180–1181. The property was subsequently sold to a shoemaker at a municipal auction in 1809. SSA, Stockholms magistrat och rådhusrätt, A6a, Lagfarts- uppbudsprotokoll, vol. 52 (1807–1809), 1809, pp. 159, 188, 207.

52 For the introduction of the Jacquard loom to Sweden, see Ciszuk (2013), p. 91. For the Swedish textile industry during the nineteenth century, see Schön (1979).

53 SSA, Stockholms magistrat och rådhusrätt, F6a, Konkursakter, huvudserie, 155/1816.

54 SSA, Stockholms magistrat och rådhusrätt, F6a, Konkursakter, huvudserie, 141/1798.

55 SSA, Stockholms magistrat och rådhusrätt, F6a, Konkursakter, huvudserie, 68/1810.

7 Hair professionals in financial distress in Stockholm, 1750–1830

Riina Turunen and Kustaa H. J. Vilkuna

Introduction

Hair and hairdressing was an important societal, political and fashion-related matter throughout early modern Europe. Using a wig or styling natural hair to look like a wig was an inseparable piece of an outfit for members of high society by the mid-seventeenth century. For members of the elite, using a wig was a way to show superior social rank and political power and to separate themselves from common people, who did not wear wigs.[1]

In the eighteenth century, the social and political implications of hairdressing started to reflect the development of a new consumerism. This was characterized by a middle class, which began to emulate the consuming habits of the higher social stratum in order to show their position in the societal hierarchy. In this, purchasable luxurious products, such as colonial import goods, as well as fashionable outfits, including wigs and dressed hair, were important elements.[2] By the mid-eighteenth century, nearly every male head in the high and middle social strata across much of Europe was adorned with a wig.[3]

When the formerly high-end luxury product developed into a more widespread luxurious consumer product, the need for professionals who could manufacture, maintain and decorate wigs, or who could fashion natural hair to resemble the look of wigs, rose.[4] The actual wigs were not that expensive after shorter and smaller wigs came into fashion from the mid-eighteenth century onwards. Maintaining wigs and styling natural hair fashionably required visits to a wigmaker or a coiffeur, however, typically many times a week. This made hair services a luxury, which was not available to everyone.[5]

The growing demand and the social and political significance of hair made hair professionals important members of early modern societies. This provided them, at least in theory, with a position where they could earn a decent living and even accumulate wealth. Nevertheless, many hair professionals had repayment problems, faced poverty and ended up insolvent. In this chapter, we will discuss how hair professionals in Stockholm managed these types of situations between 1750 and 1830.

Focus of the investigation

A number of changes that affected the hair business as an industrial field occurred during the period in question. Wigs were a commodity and hair business a type of handicraft that prospered and declined according to fashion and taste. Wigs started to go out of fashion due to political, medical and gender-related reasons in the late eighteenth century.[6] In the next century, they disappeared from everyday use and in Swedish society for the most part only appeared on the stages of theatres. The end of the early modern hair business industry not only meant that the demand for hair services decreased but also that the once needed and valued role of hair professionals largely was forgotten.[7]

We will argue that there were also other causes for the difficult financial situation that faced Stockholm hair professionals than the weakening of the industrial field. These include causes that relate to the access to credit markets and to the idiosyncrasies of the hairdressing industry. The aim is to present novel insights into the interplay between luxury, fashion, credit markets and financial troubles in the early modern period.

Natascha Coquery has analysed the management of uncertainty by bankrupt Parisian jewellers and tapestry makers in the eighteenth century and noted that the shops of these luxury product and service producers functioned as credit transfer hubs, where the shopkeepers worked between those who owed them and those to whom they owed money.[8] Stockholm wigmakers and coiffeurs fit this description. They both lent and borrowed money; a bidirectional position in the credit market made them vulnerable to the inherent risks of both these activities. For this reason, we view the Stockholm hair professionals as an excellent case study.

Two further perspectives will be used to complement the existing body of research. The research on early modern credit markets and insolvency is substantial, but has mostly focused on socially quite restricted groups. The activity of merchants, aristocrats and peasants in credit markets is well documented,[9] and insolvent merchants and trading firms have been widely analysed.[10] In studies on bankruptcy, the focus on people and firms with a connection to trade likely can be explained by the simple fact that in certain geographical areas, bankruptcy mostly affected these groups.[11] In many early modern states, only merchants were allowed to file for bankruptcy.[12]

None of this holds true in early modern Stockholm, where the bankruptcy option was open to everyone, and where merchants only formed a small share of all bankrupt debtors.[13] By focusing on the experience of a little researched group such as hair professionals, a socially more varied perspective, and a better understanding of how financial difficulties related to the access to credit markets were handled, will be secured. This is in line with the emphasis placed by Laurence Fontaine who views the early modern credit market as a sectionalized entity, where the cultures, practices, purposes and ways of lending and borrowing were distinctive between different social groups.[14]

In addition to the socially more varied view, we also aim to produce a more multi-voiced perspective. This will be accomplished by letting the wigmakers and coiffeurs have their say through an analysis of bankruptcy application letters: the written statements of cause and effect and responsibilities that were handed in at the initiation of a bankruptcy process. Even though a plethora of causes for insolvency has been mentioned and proposed in various studies, the actual viewpoints of the debtors have only rarely been included in the analyses.[15]

In order to understand why Stockholm hair professionals frequently ended up in financial troubles, we both discuss individuals from the group who declared bankruptcy as well as impoverished but ultimately solvent individuals. According to the online database *Tidigmoderna Konkurser*, 36 bankruptcy processes that included wigmakers and coiffeurs were initiated between 1750 and 1830 at the Stockholm Town Hall Court. As some wigmakers and coiffeurs filed for bankruptcy more than once, the total number of individual debtors was 28.[16]

Even though the share of hair business professionals of all bankruptcies that were initiated during this period (8,600) was small, it still shows that hair professionals belonged to a group of people in early modern society, who were involved in formalized credit relationships that could lead to bankruptcy. This can be contrasted for example with verbally agreed debt liabilities, a type of obligation that would not lead to bankruptcy proceedings.[17] Bankruptcy cases were still only the tip of an iceberg, however, as there were numerous hair professionals who ended up in poverty but never filed for bankruptcy.

The empirical analysis is based on a wide number of original sources. The basis for the analysis is a comprehensive, still unpublished database that contains thousands of occupational biographies of wigmaker masters and journeymen and coiffeurs in Sweden in the period 1648–1810. The database has been compiled from various original sources, including tax registers, church records and newspapers. It also includes data from the probate inventories of deceased hair professionals and their wives as well as information from bankruptcy files from the Town Hall Court. Bankruptcy files are available in 24 out of the 36 cases. Out of these, we were able to analyse 18 cases that belonged to 15 individual debtors (see Table 7.1).

In the case of the bankruptcy files, special attention has been given to the bankruptcy application letters where the debtors gave their reasoning for their insolvency. The wording of these letters was affected by various contextual factors, including the expectations of the court, and the framework set out by the bankruptcy laws. Still, only a debtor knew exactly what had happened and what he or she saw as the decisive factors behind the financial failure.[18]

Table 7.1 Bankruptcies initiated by wigmakers and coiffeurs at the Stockholm Town Hall Court, 1750–1830

Date of filing	Title	Name	Bankruptcy file available	Bankruptcy file used in this chapter
11 Apr 1763	Wigmaker	Jean Coquet	No	No
7 Nov 1761	Wigmaker	Anders Slottberg	No	No
27 Feb 1767	Wigmaker	Jakob Wessman	No	No
10 Jul 1767	Wigmaker	Jakob Westman	No	No
4 Sep 1769	Coiffeur	Peter Hedman	No	No
8 Dec 1771	Wigmaker	Anders Slottsberg	No	No
15 Feb 1772	The oldest of the wigmakers	Hindric Rodecor	No	No
11 Dec 1771	Wigmaker	Anders Slottberg	No	No
15 Feb 1773	The oldest of the wigmakers	Hind. Rodecor	No	No
26 Apr 1777	Wigmaker	Sven Klint	Yes	Yes
7 Nov 1778	Wigmaker	And. Slottberg	Yes	Yes
23 Jul 1781	Wigmaker	Amb. Molander	Yes	Yes
14 Sep 1782	Wigmaker	Anders Risberg	Yes	Yes
13 Aug 1785	Wigmaker journeyman	Carl Hellman	Yes	Yes
27 Sep 1786	Coiffeur	Nils H. Söderling	Yes	No
23 Apr 1787	Coiffeur	Jonas Olander	Yes	No
2 Nov 1787	Wigmaker	Johan Gotjer	No	No
15 Nov 1787	Coiffeur	Pier Brette	Yes	No
17 Jan 1788	Wigmaker	Joh. Lundberg	Yes	Yes
17 Jun 1793	Wigmaker	Johan Rising	No	No
11 Jun 1795	Wigmaker journeyman	Adrian A. Biörkström	No	No
25 May 1797	Wigmaker	Abraham Lönngren	Yes	Yes
13 Jun 1797	Wigmaker	Nathanaël Sommar	Yes	Yes
2 Oct 1797	Wigmaker	Johan Bohlin	Yes	Yes
16 Jan 1798	Coiffeur	Peter Dahlberg	Yes	No
16 May 1798	Wigmaker	Peter Törnquist	Yes	Yes
15 Apr 1799	Wigmaker	Abraham Lönngren	Yes	Yes
3 May 1799	Coiffeur	Gaspard Chartés	Yes	No
19 Nov 1801	Wigmaker	Johan Göthelius	Yes	Yes
8 Feb 1806	Coiffeur	Carl G. Wickström	Yes	No
15 Sep 1806	Coiffeur	Gaspard Charles	Yes	Yes
9 May 1807	Wigmaker	J. Lundberg	Yes	Yes
4 Nov 1807	Wigmaker journeyman	Otto R. Engström	Yes	Yes
23 May 1808	Coiffeur	Gaspard Charles	Yes	Yes
1 Oct 1821	Wigmaker	Petter Hultgren	Yes	Yes
19 Sep 1828	Wigmaker	Petter Hultgren	Yes	Yes

Source: www.tidigmodernakonkurser.se.

Changes in the socio-cultural system, fashion and taste

As mentioned in the introduction, the hair industry was a business sector that depended on fashion and taste for its survival or demise. If wearing wigs and curling hair were no longer favoured, the hair industry could not thrive. As it happened, changes in fashion, taste and the socio-cultural system caused a weakened demand for hair products and services in the late eighteenth century which proved damaging for the Stockholm hair industry.

Wigs and the curling of hair did not become unfashionable overnight; nor did the demand for cutting and styling hair disappear entirely. Traditional hair professionals, that is, educated and trained master wigmakers and coiffeurs still felt the gradual change in fashion towards the end of the eighteenth century. At first, the symbol of Old Regime Europe, long full-bottom wigs were replaced by shorter and neater styles, and finally, wigs were abandoned entirely.[19]

In the 1770s, bankrupt wigmakers and coiffeurs did not worry about their future too much. Their bankruptcy application letters instead mention short-term aspects such as financial distress caused by fluctuating economic cycles as well as changes in the prices of raw materials, work force and finished products.[20] This included a deflation of the price for finished wigs, which can be corroborated from estate inventories.

After the 1770s, the tone in the bankruptcy application letters turned darker, and the future prospects for hair professionals started to look grimmer. In the last decades of the eighteenth century, only a few Stockholm wigmakers operated profitable businesses. According to the hair professionals who went bankrupt, the financial problems were typically caused by uncertainties in the wider hairdressing industry, declining profitability and collapses of individual businesses. In 1788, for example, a bankrupt wigmaker master noted that his customers no longer bought new wigs.[21]

By the 1790s at the latest, the gloomy future for the hair business was well recognized. Two bankrupt masters expressed this clearly in 1798. One of them described how his poor financial standing was caused by a breakdown of the whole wig-making industry. The other one explained how wigs were going out of fashion and thus how the handicraft that used human hair as its raw material had totally collapsed.[22] Other bankrupt hair professionals lamented how the wigmaker's trade had been in decline for many years and continued to perform badly, with the consequence that the manufacture of wigs and the styling and curling of hair no longer were viable businesses.[23] The availability of raw material had worsened, and working opportunities and income had significantly lessened.[24]

The end of the hair business industrial life cycle affected the master wigmakers in particular. It was socially degrading for a master wigmaker to start working for someone else or for instance to accept work as a coiffeur. Thus they tried to maintain what was left of their previous social and economic position even if this led to impoverishment. Shifting work or moving to a completely different line of business was not as problematic for journeymen and coiffeurs. The evidence at hand suggests that they often abandoned their previous positions and sought employment among the gentry, applied for jobs as valets or lackeys or joined the army.

The structural change that the hair business as an industrial field went through led many hair professionals to engage in supplementary work to compensate for their diminishing incomes. Many ended up speculating in real estate or entered into the restaurant business.[25] It was relatively common for the wives of hair professionals to engage in smaller trading businesses or to run a pub under their husbands' name.[26]

Gaspard Charles, the coiffeur to the Queen, supplemented his income by importing female hairpieces, and his wife sold knick-knacks. The wife of the wigmaker journeyman Otto Reinhold Engström was a seamstress whereas the wife of the wigmaker master Johan Lundberg invested in real estate.[27] Only rarely did the supplementary work become profitable, however. It was more common that the secondary activities forced the hair professionals to obtain new credits, negatively impacted their ability to pay taxes, led to temporary incarceration in debtors' prison or even to bankruptcies.[28]

Troublesome clients and the habit of non-payment

Even though the end of the industrial lifecycle was a key reason why wigmakers and coiffeurs lost their socio-economic position in the late eighteenth century, the double-edged sword of credit also played a part. Access to credit markets was needed in order to do business in the first place. Bad credit at the same time would prevent someone from achieving a balanced business.

Buying and selling on credit was the typical way in which early modern markets worked, as ready money was in short supply. Stockholm hair professionals made purchases on credit too. The bankruptcy documents for instance show that they regularly bought fabrics, clothes, foodstuff and spices on credit. This normal behaviour could become dangerous, when combined with the workings of the credit markets. For hair professionals, selling on credit was especially problematic because of their clientele, which mainly consisted of wealthy noblemen and officials, who were especially careless with repayments. As these customers typically used hair services many times a week their pile of unpaid bills grew fast. As a result, agreements for annual or monthly payments were normally made. This did not mean that payments were made on time, however.[29]

A wigmaker master who filed for bankruptcy in 1798 complained that he had suffered for years from his unreliable customers, some of whom had died before repaying their debts. Similarly, he believed that some of his still-living clients would never repay him. His clients included two Stockholm town secretaries, both noblemen, one of whom had used the wigmaker's services for years but had never repaid his debts.[30]

Carl Hellman, who made his living as a coiffeur, did not end up bankrupt because of mismanagement of his own finance but because of his negligent customers. His clientele was made up of secretaries and officers with annual contracts, who did not bother much with repayments. Hellman's assets at the beginning of his bankruptcy proceedings in 1786, as a result, consisted almost entirely of receivables from his customers. Hellman at the same time was meticulous with his own finances and had paid his own shopping with bills that corresponded to what he was owned by his customers.[31]

The most unreliable customers were those connected to the Royal Court. Johan Olof Beckmansson started his career as a coiffeur for the pages at the Court in the 1790s and later became a coiffeur for the Queen dowager. He would teeter on the brink of bankruptcy in 1802, largely because of chronically late repayments from his clients. His wife also worked for the Court, as a lady coiffeur, and as the salaries paid by the Court were chronically late, her clientele was likely similarly unreliable as her husband's.[32]

The Queen's coiffeur Gaspard Charles not only took care of the Queen's hair and styled and curled the hair of courtiers and court maids but also delivered fashionable hairpieces to the court ladies. The court employees typically did not bother to pay off their purchases or sometimes just moved away, leaving Charles unaware from whom he should demand payment. This had a role in the three bankruptcies that he went through.[33]

The multiple bankruptcies at the same time show that Charles bounced back after going bankrupt, as he clearly was able to obtain new credit between these occasions. As Mats Hayen has shown, the renewal of the Swedish bankruptcy legislation between 1766 and 1773 changed how a bankruptcy was experienced by a debtor, by offering protection, helping to reorganize finances and generally provide a way out of a hopeless situation.[34]

Hair professionals had to accept a repetitive habit of non-payment from their customers. In general, this was how the early modern economy functioned: everyone was indebted to someone else and repayment took place irregularly or not at all.[35] Allowing purchases on credit and extending repayments were necessary in order to maintain customer relations and profit in the long run.[36] As has been proven to be the case for merchants, it was probably also an important advantage and a marketing tool for artisans to be able to offer better credit terms to their customers than their rivals.[37]

The nature of the debts must also be taken into account. Debts owed by hair business customers were typically registered in the form of an outstanding balance in an account book. In early modern Sweden, account books were seen as legal proof of a credit relationship, but only if the book was kept by a merchant.[38] The account books of artisans did not carry the same weight as evidence in a court of law. As a result there were not many legal possibilities for hair professionals to claim for recovery of unpaid bills for unpaid services and purchases from their customers. Even applying to make a customer bankrupt was not always an effective option because bankruptcy litigations could last for many years, even decades.[39]

The customers did not repay the money for the services and products that they bought from the hair professionals for a number of reasons. Many customers lacked sufficient capital to do so. In the eighteenth century, a post or a job did not mean a regular salary. Even when paid on time, salaries were typically not enough to cover the living expenses for most officials, irrespective of their social rank.[40] If an official lacked other sources of income, it was hard to maintain a satisfactory lifestyle without running into debts. The situation was not unique to Sweden. Throughout Europe, aristocrats, in particular, were heavily indebted because of dowries, extravagant lifestyles and requirements connected to a life in the public eye.[41]

The successive circles of credit, an idea presented by Laurence Fontaine, likely played a role, too. According to Fontaine, credit was obtained in a certain order, first from family, friends and neighbours and only after this, from unfamiliar people and institutions.[42] When repaying the debt, the direction was reversed, with the closest and most familiar creditors first in line to be repaid. Following this idea, it was not likely that hair professionals were first in line when their clients repaid their debts.

An expensive profession

Hair business professionals often ended up in financial troubles and accumulated debts because of the inherent costliness of their line of work. What was necessary for the operation of their businesses was also frequently financially burdening. Three aspects in particular affected this situation.

First, living costs often made up a disproportionately large part of their expenses. The residence and workshop of a hair professional was typically located in the same building or apartment. This, in turn, had to be situated where the customers lived. Because the demand for wigs and hair services was highest in the more affluent areas of the city, usually this meant that hair professionals also lived and operated their businesses in the same areas. This included the Old Town, and Norrmalm, the area just north of the city centre, where a successful wigmaker master potentially could reach at least 400 customers. In Ladugårdslandet, on the north eastern outskirts of the capital, the number of potential customers in contrast was far fewer. For many hair professionals, this meant that they ended up living too expensively, beyond their means. The wigmaker master Sven Klint, for example, owned one-fourth in a house on the outskirts of Stockholm. Instead of living and carrying out his business there, he rented a room in the heart of the city.[43] In affluent areas like this, the yearly rent for a small apartment was higher than the purchase price of a house in a smaller Swedish town.[44]

Second, hair professionals frequently were forced to emulate the lifestyles of their customers in order to attract new clients on a competitive market. In this, they turned to luxurious consumption, which was often paid on credit or financed at the expense of assets that were needed to run their businesses. By investing in outfits and nicely decorated apartments, hair professionals became not only just the manufactures but also the consumers of luxury products and services. It is hard to say how much it was because of this conscious consumption made in order to lure customers or because of other reasons that hair professionals eventually ended up in bankruptcy. We can assume that, in many cases, it made their financial burden heavier, however.

The behaviour can be seen in the life of Abraham Lönngrén. Lönngrén came from humble beginnings in Åbo in the Finnish part of the Swedish Empire. After moving to Stockholm, he became a successful hair business professional and, by the 1790s, owned one of the largest wig-making businesses in the capital. By this point, his lifestyle was far from modest. When Lönngrén walked around the streets of Stockholm, he wore several golden rings on his fingers and carried a walking stick with a golden knob under his arm. He could check the passing of time on his gold watch and set the table at his home with silver dishes, fruit baskets and different sets of fine cups and glasses to serve coffee, tea or wine. His personal seal was golden. The family's numerous pieces of furniture were fashionably colourful. Yet, when Lönngrén ended up in bankruptcy, he complained bitterly how wealthier people envied a person like him, who was hard working and industrious. This was why, according to Lönngrén, they took away any chances from him to make a living.[45]

Lönngrén did not elaborate why the wealthier people would envy him or try to prevent him from making a living. His outburst might reveal, however, that he recognized that despite his efforts to try to live like a wealthy gentleman, he was and never would be part of real high society. In a way, Lönngrén had tried to live by the rules of a bygone world. In the first half of the eighteenth century, the wigmaker masters had belonged to the urban elites. After this, this kind of societal position was possible only if a wigmaker master succeeded in real estate business, for example.

The wigmaker master Wilhelm Meyerholtz also spent good sums of money to entertain clients and to look like a successful gentleman. Like many young journeymen, he married a wigmaker widow in 1751 and received burgher rights. When his wife died, he inherited her substantial assets. His second marriage to the widow of a military official added new assets to his estate. For many years, Meyerholtz operated a wig-making workshop and was able to employ at least two journeymen. Gradually, his finances started to dwindle, and by the time of his death, at the age of 31, in 1759, he left an insolvent estate to his wife and children. The money had been spent for instance on stylish suits, a silk cape, silk socks, a tea set, tools for taking snuff and modern furniture.[46]

Conspicuous consumption was not restricted to the wigmaker masters. Coiffeurs and journeymen – especially if they served the gentry as valets – also shared the desire and need for a luxurious lifestyle and imitated the look and behaviour of their employers. This included owning hunting dogs as well as engaging in gambling.[47] They also had a taste for good food and drinks. The coiffeur to the Queen, Gaspard Charles, and his wife, for example, spent an amount that was equivalent to the annual salary for their housemaid on coffee, sugar, syrup, raisins, vinegar and many kinds of expensive seasonings in little more than two months at the turn of 1808.[48]

The desire to imitate the lifestyle of the upper classes was typical, not just for wigmakers and coiffeurs, but for wealthy burghers in general. In the early modern period, luxurious consumption was important in order to show one's societal position. A person's character was judged by his appearance.[49] When hair professionals tried to emulate the ways of high society, it was done in order to highlight their professionalism as hair service producers. This was not always a financially sound behaviour, however. While it was good for business to look fashionable, refined and to smell good, when circumstances turned sour, money spent on silk socks and fashionable furniture was needed elsewhere.

A third financial burden was costs associated with the guild system to which hair professionals were bound. The largest costs were those incurred when trying to advance in rank. In order to obtain rights as a master, a wigmaker journeyman had to pass an examination before The Wigmaker's Guild. This was paid by the candidate, who had to buy raw materials, rent a workshop and pay various fees to the guild and to the state. After a master examination was accepted, the new master had to tailor a suit that was worthy of his new status, send inquiries to find customers, pay burgher fees and find an apartment and a workshop. At this point in life, he was also likely to get married. In sum, by the time a young man had advanced in his career to a point where he received burgher rights as a master, he had incurred substantial costs and was as a result also often heavily indebted.[50]

After entering business, the real work began. A hair professional had to work industriously in order to build up a large and returning clientele. Anders Risberg obtained his burgher rights as a wigmaker master in 1778 and gradually increased the number of employees in his workshop. In 1783, when he filed for bankruptcy, he employed three journeymen and an apprentice. This was not enough to become successful, however. In his bankruptcy petition letter, Risberg explained how he had gone insolvent because he had not been able to secure a sufficient number of customers during four years of operation, to be able to repay his creditors.[51]

When the wigmaker journeyman Johan Lundberg filed for bankruptcy in 1788, he angrily explained to the court that the cause of his misfortune was the Swedish judicial apparatus and guild system. Lundberg explained how he had tried to save the penniless widow of a wigmaker and her four children from distress by marrying her and taking over the operation of the workshop that she had inherited. He had subsequently used his scant resources for a master's examination. The Wigmaker's Guild had rejected the examination, however. To add to his misery, the guild subsequently visited Lundberg's workshop and confiscated the signboard that advertised his business. The move caused the loss of customers, and the only source of subsequent income was Lundberg's landlady, whose hair he regularly styled and curled. Lundberg took the matter to the Town Hall Court, but the case was left resting for almost two years.[52]

Inherited debts

The strategy of remarrying for subsistence and the role of inherited debts for business have rarely been discussed in studies about early modern insolvency, even though debt in particular clearly could darken the premises for future life and make it hard to find balance in business.

Anders Slottberg, a wigmaker master who had trained in Paris, started his career in Stockholm without any financial support from his family.[53] His first marriage, to Anna Sabina Sachtleben, the daughter of a humble German wigmaker master did little to help his financial situation. Sachtleben died not long after the marriage, and Slottberg soon thereafter married again. His second wife was Charlotta Lovisa Schubardt, a dancer at the Royal Opera. She was the daughter of Salomon Schubardt, a wigmaker master with wide-ranging interests in the hair industry but also a distinct taste for luxury. His business interests and lifestyle impacted his finances negatively, and he died poor and bankrupt. Slottberg had already before this point taken Schubardt's workshop under his care. When his father-in-law died, he and his wife inherited nothing more than debts.

These financial liabilities, combined with illness and the cost associated with the burial of several of their children caused Slottberg and his wife to file for bankruptcy in 1771. The bankruptcy included his two sisters-in-law, who also worked as dancers at the Royal Opera. In an attempt to find a way out of the situation, Slottberg was initially granted an exemption from tax payments but was in the end declared broke by three other wigmaker masters. Some of the creditors subsequently approved an agreement whereby the debtors would be allowed to repay their debts from the salaries that were forthcoming from the Royal Opera. After a couple of years, this agreement was abandoned, possibly because the repayments stopped. A new bankruptcy followed in 1778. According to Slottberg, this was again caused by ill health, but also by the lack of means of livelihood as well as the high costs of living. Much like his customers, Slottberg also owed money to other small shopkeepers after having purchased things like fabrics, food and brushes on credit, an indication that bankruptcy not necessarily ruled out a burgher from the short-term credit market.[54]

Marrying the widow of a master was a typical strategy for a journeyman to secure burgher rights in the early modern period. For the widow, this was a way to get a male head of the household who could invest in and continue the operation of a workshop. If the widow was poor, this could lead to financial distress for the journeyman, however. This was common in the Stockholm hair professional community.

The situation can be identified in the life of the previously mentioned wigmaker Johan Lundberg. Lundberg's career started out in 1779 when he became employed as a journeyman in a wigmaker workshop ran by Ulrika Neuman, founded by her late husband, the wigmaker master Carl Biörk. Neuman was not allowed to file for bankruptcy after the passing of her husband in 1775. Even compiling an estate inventory was left undone for years.[55] When it was drawn up in 1782, it showed that Neuman possessed a range of tangible assets linked to wig making. This included the workshop, with some tools and the signboard that proved that it was the workshop of an official wigmaker. Among the intangible assets was the important legal right to manufacture wigs.

As the daughter of a wig master, Neuman most certainly knew how to manufacture wigs and how to curl and style hair even before she married Biörk. A woman in charge did not make a workshop reliable in the eyes of customers, however. A man was needed, and soon journeymen were hired. A number of different journeymen were employed but left after arguing with Neuman and after failing to see any future in the nearly bankrupt business. Only after Lundberg was hired as a journeyman, did things change. When the two wed in 1782, Lundberg took over the workshop and accepted Neuman and her children under his guardianship. Similarly, he assumed liability for all debts that Neuman had inherited.

The situation was not long-term successful. In 1788, Lundberg and Neuman were forced to file for bankruptcy. The assets and liabilities of their estate quite tellingly matched the assets and liabilities in the estate that Neuman had inherited after her first husband 13 years earlier, in a way which suggests that Neuman and Lundberg had been unable to substantially improve their living standards.[56]

When Neuman died in 1793, the financial standing of the couple was similar to the situation five years earlier. Lundberg had to obtain credit in order to bury his wife. For Lundberg, better days were ahead, though. He married for the third time with Catharina Frisk, the daughter of a hat maker. The arrangement helped to improve his economy and saw him joining the ranks of the more successful wigmakers in Stockholm. The financial upturn did not last long. Burdened by unsuccessful ventures into real estate, the liabilities from an unprofitable pub run by his wife and the failure of a joint business undertaking with a business partner who also ended up insolvent, Lundberg in 1807 again filed for bankruptcy.[57]

The same general strategies and circumstances that affected Johan Lundberg can also be seen in the life of the wigmaker Johan Bolin. Bolin started his career in Gothenburg. In 1782, he instead moved to Stockholm, where he became employed as a journeyman in the oldest workshop in the city, originally established by the wigmaker master Jacob Isak Wessman.

Wessman had filed for bankruptcy in 1767, partly because of unsuccessful real estate investments. He would continue the operation of the workshop, even though he regularly left taxes unpaid. When he died, he left some property, the wig-making business but also substantial debts to his widow Dorothea Biurberg and their children. Biurberg, much like Ulrika Neuman, and many other widows who inherited debts after the passing of their husbands tried to find ways to continue the businesses in order to pay back the creditors.

When Bolin was employed, he invested a significant sum in the workshop, with hopes to get the money back with interest after a couple of years. This wish was not fulfilled. The money was instead used to repay the existing debts and for the upkeeping of Biurberg and her children. In the end, in 1797, Bolin was instead forced to file for bankruptcy in his own name.[58]

Conclusions

The aim of this chapter was to add to the discussion on early modern credit markets and insolvency by addressing the interplay between luxury, fashion and credit and credit-induced financial troubles. Our way to do that was to find out why Stockholm hair professionals ended up in financial difficulties, that is, in poverty and insolvency. We analysed the hair industry, as a specific field of handicraft that was affected by socio-cultural and political changes, in which fashion and the spread of luxury consumption played key roles.

Four main reasons for impoverishment and bankruptcies were proposed and discussed. These were changes in fashion, taste and in the socio-cultural system; troublesome clients; the inherent expenses of the hair business and the role played by inherited debts. Few examples of an entire industrial field that collapsed due to changes in fashion and taste can be found in the early modern period. This alone made the hair business sector an important subject of research.

Even though fashion, taste and new consumer behaviours determined a great deal of the rise and decline of the hair business, the industrial lifecycle model still does not fully explain why wigmakers and coiffeurs faced financial hardship in Stockholm.

We have instead shown that such problems often had their origins in the fact that hair professionals acted as transfer hubs on the credit market by both engaging in lending and borrowing on credit. This made them especially vulnerable to the inherent risks in the credit system. However, it was not only hair professionals who suffered from the force to sell on credit or from inherited debts, for example. These troubles touched many other market oriented people in the early modern period.

Another important result of the investigation relates to the way in which hair professionals worked to resemble their socially elevated customers in appearance and manners in order to remain competitive on a crowded market. While hair professionals emulated the consuming habits of the higher social stratum and ideally might look and smell like true gentlemen, they never fully equalled their clients. This was made apparent when their high society clients made few efforts to pay their bills in time, in a way which signalled the continued existence of social hierarchies. Whereas embracing conspicuous consumption in order to attract customers and accepting limited repayments from their clients helped hair professionals to secure business, it could also drive them into financial troubles, which, in turn, highlights how social boundaries still contributed to financial difficulties.

Notes

1 Kwass (2006), pp. 634–635, 637, 639, 644–652; Vilkuna (2019), pp. 88, 95, 97.
2 McKendrick et al. (1985); Berg (2005), pp. 5–13, 21.
3 Before the mid-eighteenth century, women used a type of tiara-like hair piece (*fontang*) made of hair. After this wigs became an almost exclusive male accessory. Vilkuna (2019), pp. 90–91.
4 Kwass (2006), pp. 635–637; Vilkuna (2019), pp. 91–92.
5 Vilkuna (2019), pp. 98, 102–104, 107.
6 Kwass (2006), p. 659.
7 In Nyström et al. (1989), 38 Swedish handicraft professions are presented in detail, but without any reference to wigmakers.
8 Coquery (2013), p. 52.
9 Muldrew (1998); Smail (2005); Hoffman et al. (2000); Ogilvie et al. (2012); Fontaine (2014); Dermineur (2019).
10 Duffy (1980); Hoppit (1987); Häberlein (2013); Nyberg and Jakobsson (2013).
11 Lee Thompson (2004), p. 88; Turunen (2017), p. 83.
12 Warren (1935), pp. 6–7; Hoppit (1987), p. 35.
13 Hayen and Nyberg (2017), p. 44.
14 Fontaine (2014).
15 For examples of the former, see Hoppit (1987), pp. 113, 126–130; Mathias (2000), pp. 15–28; Napolitano et al. (2015); Gratzer (2008), pp. 8–9. For exceptions, see Mann (2002); Reynard (2001); Turunen (2017). See also the articles in Nyberg (2017) where the bankruptcy petition letters of some single debtors have been narrated.
16 The database at: www.tidigmodernakonkurser.se. The following search terms were used: *perukmakare*; *hårfrisör*; *perukmakareålderman*; *perukmakaregesäll*; *frisör*.
17 For verbal liabilities in nineteenth-century Finland, see Turunen (2017), pp. 123–127.
18 Turunen (2017), pp. 53–54.
19 Kwass (2006), pp. 645–646.
20 SSA, Stockholms magistrat och rådhusrätt, F6a, Konkursakter, huvudserie, 102/1778; 26/1779.
21 SSA, Stockholms magistrat och rådhusrätt, F6a, Konkursakter, huvudserie, 10/1789.
22 SSA, Stockholms magistrat och rådhusrätt, F6a, Konkursakter, huvudserie, 8/1799; 124/1809.
23 SSA, Stockholms magistrat och rådhusrätt, F6a, Konkursakter, huvudserie, 112/1798; 135/1800; 197/1802.
24 SSA, Stockholms magistrat och rådhusrätt, F6a, Konkursakter, huvudserie, 207/1822.
25 *Ruotsin peruukintekijät 1648–1810*, 'De Lang, Nicolaus Louis', 'Forsberg, Sven', 'Frestare, Friedrich Gustafsson', 'Goets, Lovis Joseph', 'Granberg, Anders', 'Halle, Johan Gottlieb', 'Reimers, Christian Daniel', 'Rodecor, Heinrich', 'Uppström, Petter', 'Wahlgren, Anders', 'Westerholm, Jonas', 'Wihrström, Anders' and 'Zimmerman, Petter Ulric'.

134 *Riina Turunen and Kustaa H. J. Vilkuna*

26 *Ruotsin peruukintekijät 1648–1810*, 'Almström, Maria Christina', 'Biörck, Maria Christina', 'Bymark, Inga Maria', 'Kenckel, Maria Sophia', 'Meyer, Lisa', 'Thorberg, Maria Elisabet', 'Thunberg, Catharina', 'Törnqvist, Anna Catharina' and 'Westrin, Johanna Beata'.

27 SSA, Stockholms magistrat och rådhusrätt, Konkursakter, huvudserie, 53/1808; 23/1809.

28 *Ruotsin peruukintekijät 1648–1810*, 'Elfström, Johan Petter'; 'Rehfeldt, Johan Christian von'.

29 SSA, Stockholms magistrat och rådhusrätt, F6a, Konkursakter, huvudserie, 112/1798.

30 SSA, Stockholms magistrat och rådhusrätt, F6a, Konkursakter, huvudserie, 8/1799.

31 SSA, Stockholms magistrat och rådhusrätt, F6a, Konkursakter, huvudserie, 46/1786.

32 This was exasperated by his relatively low wage, which in 1803 amounted to forty rixdollars. *Ruotsin peruukintekijät 1648–1810*, 'Johan Beckmansson'.

33 SSA, Stockholms magistrat och rådhusrätt, F6a, Konkursakter, huvudserie, 23/1809; *Ruotsin peruukintekijät 1648–1810*, 'Charlés, Gaspard'.

34 Hayen (2017).

35 Muldrew (1998), p. 175; Fontaine (2014), pp. 52–53.

36 For example, Craig Muldrew has discussed about the reasons why so many debts were forgiven or extended in the early modern period, see Muldrew 1998, pp. 176, 181–182, 304–306.

37 For merchant credits, see Smail (2005), p. 444.

38 Grandell (1944), pp. 225–228; Turunen (2018), p. 4.

39 Nyberg and Jakobsson (2016), p. 112.

40 Ilmakunnas (2016), pp. 16–17.

41 Fontaine (2014), pp. 75–77.

42 Fontaine (2014), pp. 26, 80–81.

43 SSA, Stockholms magistrat och rådhusrätt, F6a, Konkursakter, huvudserie, 102/1778; *Ruotsin peruukintekijät 1648–1810*, 'Klint, Sven'.

44 SSA, Stockholms magistrat och rådhusrätt, F6a, Konkursakter, huvudserie, 26/1779; 30/1783; 167/1784; 112/1798; 135/1800.

45 SSA, Stockholms magistrat och rådhusrätt, F6a, Konkursakter, huvudserie, 112/1798; 135/1800; *Ruotsin peruukintekijät 1648–1810*, 'Lönngrén, Abraham'.

46 SSA, Rådhusrättens 1:a avdelning (Justitiekollegium, Förmyndarkammaren), F1A, Bouppteckningar, vol. 146: 250–256; vol. 164: 784–787; vol. 177: 209–222; *Ruotsin peruukintekijät 1648–1810*, 'Finn, Martin', 'Grooth, Hedvig Charlote', 'Meyerholtz, Wilhelm' and 'Schram, Margareta'. About the decoration and furnishing styles of nobles in eighteenth-century Stockholm, see Ilmakunnas (2016), pp. 54–82.

47 Gambling can be established from the payment of a luxury tax for owning playing card. *Ruotsin peruukintekijät 1648–1810*, 'Engström, Otto Reinhold'.

48 SSA, Stockholms magistrat och rådhusrätt, F6a, Konkursakter, huvudserie, 23/1809.

49 Ilmakunnas (2016), pp. 61–62, 72.

50 *Ruotsin peruukintekijät 1648–1810*, 'Unger, Jonas'; Embetz Skrå För Peruque Makare Embetet... (1898).

51 SSA, Stockholms magistrat och rådhusrätt, F6a, Konkursakter, huvudserie, 167/1784; *Ruotsin peruukintekijät 1648–1810*, 'Risberg, Anders'.

52 SSA, Stockholms magistrat och rådhusrätt, F6a, Konkursakter, huvudserie, 10/1789; *Ruotsin peruukintekijät 1648–1810*, 'Lundberg, Johan' and 'Neuman, Ulrica'.

53 His father was also a wigmaker master. SSA, Rådhusrättens 1:a avdelning (Justitiekollegium, Förmyndarkammaren), F1A, Bouppteckningar, vol. 134: 567–573; *Ruotsin peruukintekijät 1648–1810*, 'Slottberg, Lorens'.

54 SSA, Stockholms magistrat och rådhusrätt, F6a, Konkursakter, huvudserie, 26/1779; SSA, Rådhusrättens 1:a avdelning (Justitiekollegium, Förmyndarkammaren), F1A, Bouppteckningar, vol. 134: 567–573; *Ruotsin peruukintekijät 1648–1810*, 'Sachtleben, Anna Sabina', 'Schubardt, Charlotta Lovisa', 'Schubardt, Salomon' and 'Slottberg, Anders'.

55 SSA, Rådhusrättens 1:a avdelning (Justitiekollegium, Förmyndarkammaren), F1A, Bouppteckningar, vol. 269: 1021–1026; *Ruotsin peruukintekijät 1648–1810*, 'Biörk, Carl'.

56 SSA, Stockholms magistrat och rådhusrätt, F6a, Konkursakter, huvudserie, 10/1789.

57 SSA, Rådhusrättens 1:a avdelning (Justitiekollegium, Förmyndarkammaren), F1A, Bouppteckningar, vol. 310: 554–557; *Ruotsin peruukintekijät 1648–1810*, 'Lundberg, Johan' and 'Neuman, Ulrica'.

58 In an ironic twist, when considering the reasons for his financial difficulties, Bolin's landlady at the time of his bankruptcy was one of Jacob Isak Wessman's daughters. SSA, Stockholms magistrat och rådhusrätt, F6a, Konkursakter, huvudserie, 1/1799; *Ruotsin peruukintekijät 1648–1810*, 'Biurberg, Brita Maria Dorothea', 'Bolin, Johan' and 'Wessman, Jacob Isak'.

8 Book printing in Stockholm, from royal privilege to market economy, 1780–1850

Mats Hayen

Introduction

The book printing industry has on several occasions throughout history been dramatically affected by new technologies. Long periods of production based on one set of technological features have been disrupted by sudden changes, which in radical ways have altered the foundations of the entire industry. One such example was the invention of the steam-powered printing press, which was introduced by Friedrich König and Andreas Bauer in London in 1811. The mechanized printing press would over time completely replace the old hand-operated printing press, which had been used by book printers all over the western world, since its introduction in the middle of the fifteenth century by Johannes Gutenberg.

The introduction of cylinder printing presses in book print shops in the early nineteenth century was nothing less than a revolution. Small handicraft-based print shops either disappeared or managed to survive in specialized niches or made the technological transformation and grew into larger industries. The process forced unsuccessful enterprises out of business and made more adaptable firms stronger.[1]

New production technologies were only one aspect of the transformation of the book printing industry, however. Equally important were the changes that occurred in the market for books and newspapers and in the existing pre-modern channels of distribution.

In this chapter, I will focus on the development of the book printing industry in Stockholm, the Swedish centre for printing. The study is based partly on an analysis of records of bankruptcies of book printers in Stockholm. Specifically, I seek to show how book printers in Stockholm responded to market demands for high-end printed publications, first during the reign of King Gustav III (r. 1771–1792), second, after the issue of the Freedom of the Press Act in 1810 and, finally, after the introduction of the machine printing press in 1833. The book printing industry in Stockholm at the end of the eighteenth century was based on royal privileges, and its members were largely protected from economic pressures on an open market. The establishment of a new book printer was in normal cases only accepted by the Royal Chancellors' Office when a member of the Stockholm Book Printers' Society died or for other reasons returned the charter of privilege.

The Freedom of the Press Act of 1810 – issued after a disastrous war with Russia which was followed by the abdication of the Swedish king and a bourgeois revolution – changed the economic situation for the book printers almost overnight. Suddenly it became possible for every person who was in possession of book printing equipment to enter the industry. How did the older generation of book printers respond to this new situation? And what strategies were adopted by all members of the book printing industry – old and new – in the early nineteenth century? It seems that the competition between book printers became especially fierce in the 1830s after the introduction of steam-powered machine presses in Stockholm. As already noted, special attention will be given to high-end products. In the book industry in Sweden at the time such products were only produced by a small number of highly skilled craftsmen in the capital.

In her studies of book printing in Stockholm Anna Maria Rimm has shown that the production of luxurious books and other expensive printed products reached a new level of sophistication in the Swedish capital towards the end of the eighteenth century. Book printers at the time tried to tap into a growing market demand for cultural products centred on the King and the Royal Court, which also had a strong influence on society at large. This was followed by a turbulent period after the death of the King in 1792, when the book printing industry was forced to seek new ways to stay in business.[2] In the early 1800s, a new market for luxurious and costly products emerged. It was not based on an economy of privileges and state-controlled production, like the culture at the Royal Court and in the nobility, especially during the reign of King Gustav III. Instead, it was part of a new and expanding market for bourgeois products that signalled wealth and cultural capital for people of the middle and upper classes.[3] I will argue that the Freedom of the Press Act played a very important role in this development.

The year 1833 is also of importance. It saw the introduction of machine printing in Stockholm, and it was also one of the first years of an intense print war between conservative and liberal newspapers in Sweden. After this, the number of book printers who filed for bankruptcy increased. Between 1780 and 1850, 23 book printers in total applied for bankruptcy protection in Stockholm. Nine of these 23 cases occurred between 1780 and 1833, which is a period of 53 years. The rest, or 14 cases, were registered between 1834 and 1846, a period of only 12 years (see Table 8.1). The data makes it apparent that the book printing industry in Stockholm was quickly transformed in the 1830s, making some firms stronger and forcing many others into bankruptcy. The 1830s saw the first severe crises of the industry since its foundations had been changed by the Freedom of the Press Act of 1810.

Table 8.1 All bankruptcies by book printers in Stockholm between 1780 and 1850

No.	Years	Debtor(s)	Number of creditors	Debt in rixdollar banco
1	1780–1783	Carl Stolpe	18	6,806
2	1783–1784	Estate of Carl Stolpe	1	6
3	1796–1799	Estate of Jac. Enbom	8	496
4	1821–1822	Creditors of Johan Imnelius	39	11,166
5	1827–1829	Ernst Adolf Ortman	21	6,066
6	1827–1829	Magnus Gabriel Lundberg	27	6,738
7	1828	Carl Fredric Wennström	8	1,044
8	1830	Estate of Johan Imnelius	1	82
9	1830–1831	Carl Abraham Pihl	8	7,224
10	1834–1836	Henric Fougt	52	38,479
11	1836–1837	Carl Magnus Carlsson	23	11,243
12	1836–1837	Estate of Henric Fougt	8	8,961
13	1836–1837	Henric Gustaf Nordström	23	18,438
14	1839–1840	Carl Roselli	25	26,270
15	1842–1843	Estate of Carl Magnus Carlsson	1	3
16	1842–1844	Johan Adam Fahlman	20	5,587
17	1842–1845	Anders Gustaf Hellstén	58	30,400
18	1843–1844	Ludvig Öberg	33	21,374
19	1843–1844	Anna Charlotta Bergh	3	126
20	1843–1844	Estate of Henric Gustaf Nordström	2	3,409
21	1845–1847	Estate of Ludvig Öberg	11	7,545
22	1846	Wilhelm Emanuel Kjellström	11	1,244
23	1846–1848	Claes Joachim Schultz	18	4,404

Source: www.tidigmodernakonkurser.se.

Beginnings: the Book Printers' Society

The Stockholm Book Printers' Society was founded in 1752 as the first association for book printers in Sweden altogether. It was not created as an ordinary guild or trade association that answered to the municipal administration, but was instead placed directly under the jurisdiction of the Royal Chancellors' Office and King Adolf Fredrik (r. 1751–1771). All six book printers who were active in Stockholm at the time automatically became members of the society from the start.[4]

The number of members grew from six to seven in 1758 when Johan Georg Lange, a book printer from Gothenburg, bought a print shop that had been inoperative since 1749.[5]

The control maintained by the Royal Chancellors' Office helps to explain the relatively low number of bankruptcies in the book printing trade before 1810. While Lange managed to establish a print shop, the number of book printers grew slowly during the latter part of the eighteenth century. During this period it was very difficult for a new entrepreneur to secure a privilege as a book printer and gain entry into the existing group of book printers.

The bankruptcy case of book printer Carl Stolpe in 1780–1783

In September 1780, the book printer Carl Stolpe applied for bankruptcy. In his application, he argued that the main reason for his decision was "a number of disasters, and in particular certain loan guarantees that had proven to be to his disadvantage". He also claimed that he owned the equipment in the print shop, which at the time consisted of three hand-printing presses.[6]

Shortly after the bankruptcy case was initiated, the court was approached by Zacharias Heijenberg who contrary to Stolpe's claim presented evidence that the printing equipment in the print shop was owned by the sea captain Pehr Fontin. The transfer of ownership had been made in the spring of 1779, after which point Stolpe proceeded to rent the equipment from Fontin, and after his death in August 1779, from his widow, who was also Heijenberg's sister.[7]

The court-records also show that Fontin had borrowed money from the wholesale merchant Georg Fredrik Diedrichsson, to be able to buy the printing equipment. This proves that it was possible for someone, like Fontin, who did not belong to the Book Printers' Society, to own printing equipment that was used by a member of the society. Economic constructions like this worked against the intentions behind the system with privileges to print issued by the King and controlled by the Royal Chancellors' Office. The King was in this particular case unaware of the fact that a person without privilege – Pehr Fontin, the owner of the printing equipment – had important economic interests in the business of Carl Stolpe, who had the King's privilege to print. The authorities usually reacted negatively in cases like this, which the following proceedings in this case show.

The bookseller Johan Christofer Holmberg's application to work as a book printer had been rejected by the Book Printers' Society and the Royal Chancellors' Office earlier that year. Holmberg owned a part in a print shop operated by Lars Wennberg, and in the winter of 1782–1783 used this shop to print under Stolpe's privilege. This was considered illegal by the authorities, however, and both Holmberg and Stolpe were banned from using the printing presses in Wennberg's print shop.[8]

When Stolpe died in February 1783, his bankruptcy was still unresolved.[9] His demise had an immediate important consequence. Before the Freedom of the Press Act of 1810 the death of a member of the Book Printers' Society was more or less the only way for a new entrepreneur to acquire a privilege as a book printer and become a member of the society. And the privilege to print was before 1810 essential for anyone who wanted to run a print shop. Without it, you had to find ways to use the privilege of another printer, as some of the examples above show. The limits regarding the total number of privileges gave the authorities important power over the book printing industry, which could be used to give economic advantages and to influence the production of printed material in the Kingdom. The issuing of privileges also created important bonds between the ruler and the printers. In this specific case, the death of Stolpe enabled Holmberg to gain entry into the

Book Printers' Society, essentially by being awarded the privilege that was now freed up from the Royal Chancellors' Office.[10]

The Royal Printer and the Fougt-family

By the end of the eighteenth century the Royal Printer Elsa Fougt had established herself as the most important book printer in Stockholm and Sweden. She had learned book printing in her youth in the print shop of her father Peter Momma, the Royal Printer from 1738 to 1772. After her father's death, she worked closely together with her husband Henric Fougt, the new Royal Printer, in his print shop. After his death in 1782, she became the director of the firm. She remained in Royal Printer for almost 30 years, until 1811, when her son took over the operation. The son would end his career as a book printer with a bankruptcy in 1834, one of the first of a number of book printers who failed in their businesses in the 1830s.[11]

Elsa Fougt was in many ways the first modern book printer in Stockholm. Both she and her husband pioneered new ways to conduct business. In the 1770s, Henric Fougt for example worked closely together with King Gustav III and the newly founded Royal Opera to promote librettos and other high-end publications printed with the French Antiqua style of typefaces. This was an important break with the past, as almost all print shops in Stockholm at that time used the German Fraktur typeface and very often relied on worn-out types that produced low-quality prints.

The need for a high-quality type foundry in Stockholm was addressed by Henric Fougt in the 1770s.[12] In the estate inventory that was registered after his death in 1782, the type foundry that he had started was valued at 1,955 rixdollars, whereas the value of his print shop was estimated at 1,884 rixdollars.[13] At the time of his death, the type foundry was not recognized by the Royal Chancellors' Office, as it lacked a royal concession. Elsa Fougt managed to acquire a privilege for the type foundry in 1783 (she already had a privilege to print as Royal Printer from 1782) and to develop production at the site. Two years later six workmen, two foreign type casters and four Swedish engravers were employed at the foundry.

In a letter written in 1782 to the Royal Courts' Office Elsa Fougt had stated her competence as a book printer and director of the operation:

> as I, during the lifetime of my husband, in all his enterprises, have been his assistant, and as I, during this time, have managed to acquire all necessary knowledge in the craft of type foundry, I find myself well prepared to continue and to bring to perfection the work that he started, with the help of the foreign craftsmen that my husband with heavy expenses have brought into this country.[14]

The type foundry that was established by Elsa Fougt played a key part in the most prestigious Swedish printing project of the late eighteenth century – the Medal-History of King Gustav III, a project modelled on the *Histoire Metallique* commissioned by the French King Louis XIV. It was never finished but would have been a complete edition in print of all official medals created in Sweden.

The Swedish King specifically requested Didot-types from the French firm Fournier in Paris for the printing of the text-part of the book. In 1785, Fougt asked the Chancellor of the Royal Court Fredrik Sparre for 5,000 rixdollars as an advance which she intended to invest in the type foundry, partly by importing the requested Didot-types and partly by further developing the operation.[15] The request should be viewed in relation to a competing proposal by the book printer Johan Georg Lange, who acted with the support of the Book Printers' Society when offering to print the Medal-History for a lower sum than Fougt and with the same Didot-types. Fougt responded to Lange's proposal in the same letter to the Chancellor of the Royal Court:

> It is not sufficient for the printing of the Histoire Metallique to own the Didot-types. Diligence, knowledge and taste are needed as well. Mr Lange is certainly a quite skilled book printer. But since his main enterprise has been the production of bibles, hymn-books, postillas and other similar articles, suitable for bookbinders and for market-trade, which do not require any knowledge in foreign languages or in belles lettres, it should prove hard for Mr Lange, at an age of over sixty years, harder than he himself expects, to begin correspondence with the producers of the types needed for this work, which will require at least two years of time, and in a language which he does not understand, for a work of this scale, an undertaking never before accomplished in Sweden.[16]

From the contents of the letter it becomes clear that Fougt considered her competence as a book printer and operator of a type foundry superior to Lange's. It is also an example of the way in which competition between individual book printers was monitored and mediated directly by the King. In spite of the letter by Fougt the King decided in 1788 that the commission should go to Lange, as he requested less money than her. Nothing, however, came out of this competition between Fougt and Lange. The Medal-History project was abandoned after the assassination of the King in 1792. Lange, in turn, died the same year. Elsa Fougt continued to fight for supremacy among the book printers in Stockholm and it is obvious that she owned one of the most prosperous print shops in the city when the eighteenth century closed.

The Freedom of the Press Act of 1810

In the years after the Swedish Revolution in 1809, when King Gustav IV Adolf (r. 1792–1809) was exiled after a military coup, and his uncle, Karl

XIII (r. 1809–1818) instead became King, many small print shops were established in Stockholm. This was not directly due to the political turmoil, but instead was caused by a new Freedom of the Press Act that was issued by the new government in March 1810. The act removed nearly all regulations affecting the establishment of print shops that were previously handled by the Royal Chancellors' Office and the Book Printers' Society. The new act suddenly made it legal for all Swedish citizens to print books, newspapers and pamphlets. In effect it changed the foundations of the entire book printing industry in Sweden.[17]

Hans-Gunnar Axberger argues that this made the Swedish printing industry free, many years prior to the abolishment of the guild system which still worked within the traditional system of privileges issued by the King. The only restriction that still applied was that new printers should notify the local government that a print shop had been established, 14 days prior to the first issue from a new printing press. The new freedom was not without limitations, however. The authorities still had the power to withdraw publications that were slanderous or constituted a threat to society. This was codified in a revised Freedom of the Press Act from 1812.[18]

The Freedom of the Press Acts of 1810 and 1812 created a situation where the book printers for the first time had to compete among themselves on capitalistic terms. Competition became fiercer and the life-span of individual print shops grew shorter.[19] Most of the new print shops that were established in Stockholm in the 1810s and 1820s were started by a new category of book printers. Unlike the previous generation, these individuals often worked both as writers, publishers and printers. Typically, they acquired one or two hand-operated printing presses and used these both for commission work and to print their own books or newspapers.

Ernst Adolf Ortman was one of these newcomers. He worked as a merchant's assistant when he started a print shop in 1820. When he at the same time applied for membership to the Book Printers' Society the book printer Carl Deleen reacted with a letter to the society. In Deleen's eyes, Ortman's position as a servant, which under Swedish law placed him under the jurisdiction of the head of his household, made him unsuitable for membership in the Book Printers' Society.[20] Deleen's opposition to Ortman as an acceptable member of the society is an example of competition between entrepreneurs within the group of book printers at this time. Well-established printers tried to fend off newcomers like Ortman, who, in the eyes of Deleen and others, produced low-quality printed material and also degraded the once so important Book Printers' Society. It had lost its most important function, as a collective of book printers, each with his or her own royal privilege, through the Freedom of the Press Act of 1810.

In this case, however, it was Deleen who lost the battle. His opponent was accepted as a member of the society. Time would however prove Deleen right. Ernst Adolf Ortman never managed to establish a financially secure business and he only managed to stay in business until 1827. He applied for bankruptcy that year, after facing numerous financial problems. There were 21 creditors demanding payment and none of them belonged to the book industry. In most bankruptcy cases involving book printers, the group of creditors included business partners but in Ortman's case there were none. His main creditor was his brother, and he had also borrowed a large sum from his wife.[21]

The first fashion magazine in Sweden

Carl Deleen was one of the most important book printers in Stockholm in the first half of the nineteenth century. He started out in 1799 and quickly established himself as one of the most qualified printers in the city. His status was confirmed when he in 1801 became the official book printer for the Swedish Academy, a position that he held until 1829.[22]

In 1818, he joined forces with the artist, writer and engraver Fredrik Boye, who in 1815 had established *Konst och Nyhets Magasin för Medborgare af alla Klasser* as the first Swedish fashion magazine.[23] Together they developed Boye's paper, literally a magazine for art and news, into one of the most elaborate products on the market.[24] Boye visualized the current fashion in women's and men's clothing in each publication in high-quality colour illustrations. Almost every issue contained a detailed and illustrated article on the current fashion in Paris (and sometimes in Vienna), for women and men, and in some issues for children as well. At first, stone printing was used, but this was not deemed satisfactory because of the low quality of the prints. In 1820, the two business partners instead changed to copper printing techniques. While this was more expensive, it produced illustrations of higher quality.[25]

The high production costs and the relatively high retail price of the magazine were often mentioned in letters to the readers that were included in the magazine. This was a crucial strategy. The economy of the magazine relied entirely on sales and was therefore heavily dependent on successful marketing. The ambition to reach all levels of society was made clear in the title, which established that the magazine was "for people of all classes".[26] There is a marked difference between this product and the books and other printed material that Elsa Fougt marketed in her shop in the 1780s. Those were products of privilege, often directly ordered by the authorities or financially secured through the control of competition, which was an important feature of the print industry before 1810. The fashion magazine produced by Boye and Deleen, in contrast, had to compete in an emerging market economy, without protection from competition and always dependent on the number of copies that could be marketed and sold each month. Elsa Fougt, in comparison, had belonged to a generation of book printers who could conduct business mainly by addressing the King and other high-ranking officials. In her eyes, the public, who ultimately bought the products, were important, but not as important as the King. In the early nineteenth century this was no longer true.

The fashion magazine of Deleen and Boye was successful; it lived on for a long time and was known in most parts of Sweden. Boye was a master engraver and his articles and pictures of Parisian fashion were very popular among women at the time.[27] Carl Deleen was also a master of his art and paid much attention to the technical education of his apprentices. His print shop was known for its high quality, and he was often chosen for important publications. Carl Deleen continued to produce high-quality publications well into the 1840s. He could not, however, compete with more efficient printing firms – like P. A. Norstedt & Sons – in his later years. Deleen was eventually forced to accept a small pension from the King Karl XIV Johan to make ends meet.[28]

The start of industrialized book printing

The first mechanized machine printing press in Sweden was installed by the book printer Nils Magnus Lindh in the city of Örebro in 1829. It was a rotating press with two cylinders from the manufacturers Augustus Applegath and Edward Cowper in London, who had successfully constructed industrial printing presses that were cheaper than the first models by König and Bauer.[29] This made the new technology more accessible to book printers in England and abroad. The reputation of the print shop in Örebro was bad, and few people at the time believed that high-quality books could be produced by the new machines. Lindh was often criticized for the low quality of his printed material. It still was one of the largest print shops in Sweden and produced catechism books in large quantities that were sold by Lindh through a network of agents in cities and at county fairs throughout Sweden.[30]

The first two machine presses in Stockholm were imported by the publisher and book printer Lars Johan Hierta in 1833. Much like in Örebro, they were built by Applegath and Cowper. The first document that records these two machine presses is an article in *Aftonbladet* published on 27 December 1834. The print shop of Lars Johan Hierta had approximately 30 employees in 1834.

The third documented industrial printing press in Stockholm was used for the first time in December 1835. It had one cylinder and came – like the previous machines in Sweden – from the workshop of Applegath and Cowper in London. It was imported to print Stats-Tidningen – the official newspaper of the Swedish government. The machines in the print shops of *Aftonbladet* (liberal) and *Stats-Tidningen* (conservative) immediately became important during the intense print wars of these times (see below).

The introduction of mechanized production methods in the book printing industry happened at the same time as another crucial development, namely the growth of a small number of print shops into large enterprises. The most important of these was the firm P. A. Norstedt & Sons.

In 1821, the merchant Per Adolf Norstedt bought a print shop and type foundry in Stockholm from the widow of the book printer Johan Petter

Lindh, who had died the previous year.[31] At the time of the acquisition, the number of hand-operated printing presses in the print shop was very high, when compared to other book printers in Stockholm, in effect making it the largest operation of its kind. After the issue of the Freedom of the Press Act in 1810 the business strategies of the book printers in Stockholm changed character. Most of the print shops that were founded after 1810 remained small-scale operations with typically only between one or two hand presses. A number of print shops, such as the one operated by Lindh, grew much larger than before, however.[32]

The development in Stockholm can be likened to the situation in London, where already by the middle of the eighteenth century a few book printers had grown their businesses into large enterprises. One example was the firm managed by Samuel Richardson who operated three print shops with more than 120 employees. Most print shops in London remained small, however, with few employees and only one or two hand-operated printing presses.[33]

In 1823 Norstedt made his two sons, Carl and Adolf, partners in the business. In 1828 the founder retired and the company continued with Carl as head of the production and Adolf as head of administration. At the time the company rested on three legs: the print shop, the type foundry and the publishing house. In 1833, the rights as Royal Printer were transferred from Henric Fougt to P. A. Norstedt & Sons. Shortly after that, Fougt applied for bankruptcy.[34]

The print wars from 1830 to 1844

The foundation of the liberal newspaper *Aftonbladet* in 1830 marked the beginning of a period of print wars in Stockholm. The main belligerents were Lars Johan Hierta, owner of *Aftonbladet*, and the Royal Librarian, Per Adam Wallmark, who since 1835, by royal decree, functioned as the publisher of the government newspaper *Stats-Tidningen*.

The Chancellor of the Royal Court in 1834 had proposed that the oldest Swedish newspaper *Post- och Inrikes-Tidningar* should be made into the political organ of the government and that the paper thereafter should be called *Stats-Tidningen* – the State-Paper. The reason was the need for a publication which the government and other conservative forces could use to counteract the strong liberal political current in the early 1830s. The Swedish Academy, which since 1791 had held the royal privilege to publish *Post- och Inrikes-Tidningar*, wanted no part in the proposal and instead offered to transfer the privilege to whomever the government choose as the new publisher. In exchange the academy wanted a yearly reimbursement from the state. This was granted by the King for the period 1835–1840.[35]

Per Adam Wallmark was a fierce opponent of the new romanticist movement in Swedish literature. Throughout his life he remained a proponent of French classicism and expressed strong conservative viewpoints on political matters. His strong ties to the Royal Court explain why he was chosen as the publisher of *Stats-Tidningen*. It was in short a political appointment directed against the liberal movement, in general, and against the liberal paper *Aftonbladet*, in particular.

During his first year as publisher Wallmark used the print shop of the book printer Anders Gustaf Hellstén. It quickly became apparent to Wallmark that this arrangement was too expensive. In the summer of 1835 the Chancellor of the Royal Court wrote a letter to the King stating that the printing costs of the paper could be significantly reduced if printing was done in a print shop controlled and owned by the state. In the end this resulted in Wallmark acquiring his own print shop. The contract with Hellstén was terminated and Wallmark imported a new rotating printing press from Applegath and Cowper to further reduce printing costs. The last issue of *Stats-Tidningen* printed on the hand-printing presses in Hellstén's print shop was published on Christmas Eve 1835. The first issue of the machine-printed paper, from Wallmark's print shop, went on sale in Stockholm four days later.[36]

Stats-Tidningen immediately faced rising production costs. Wallmark had to raise subscription fees and with the help of the government force more state institutions to become subscribers. He also tried to make the most of his monopoly on legal ads, which was not popular among his main opponents, the liberal newspapers.

The print war was eventually won by the liberal newspapers, led by *Aftonbladet* and its owner Lars Johan Hierta. *Stats-Tidningen* managed to stay afloat a couple of years but became a major financial disappointment for the government and the publisher Wallmark from 1838 onwards. In 1844, the government withdrew from the venture, and *Stats-Tidningen* reverted back to its old name.[37]

The book printers in Stockholm came under a lot of pressure in the period between 1830 and 1844. This was due to increasing competition in part connected to the print wars and in part to the introduction of the mechanized printing press. There were economic downturns on the European market as well in 1829–1832, 1837–1842 and 1847–1848.[38] In Stockholm, the first of these downturns is an important background factor explaining the need to find cheaper ways to produce printed material, as the development in the print shops of *Aftonbladet* and *Stats-Tidningen* shows. It might also have affected the very large bankruptcy of Henric Fougt in 1834. There were 51 creditors in his case and the largest total sum of debts – 38,479 rixdollars – of all book printers in this study. Exact details of his bankruptcy, have not, however, survived to this day as the case file is lost. The next downturn between 1837 and 1842 also saw many bankruptcies among the book printers in Stockholm. Carl Roselli went bankrupt in 1839, Johan Adam Fahlman did the same in 1842 and was followed later that same year by Anders Gustaf Hellstén. The bankruptcy of Hellstén was directly linked to developments regarding the production of *Stats-Tidningen*.

The early decades of the nineteenth century saw many changes in the world of books and book printing. For the first time, the production and distribution in the printing industry became dominated by economic forces, which led to a powerful influx of innovative and diverse publishing methods. It is evident when comparing the early nineteenth century with the late-eighteenth century that the state had lost control over the book printing industry and had to compete on more or less equal financial terms with other forces on the market: publishers, print firms and strong independent newspapers.[39]

The bankruptcy of Ludvig Öberg in 1843

One of the victims of these turbulent times was the book printer Ludvig Öberg, who applied for bankruptcy in January 1843. Born in 1795, Öberg started work as a journeyman in a print shop owned by Elmén and Granberg in 1822. Five years later he became its owner. The print shop mainly printed popular literature: handbooks, travel guides, as well as poetry and novels.[40] At the time of his bankruptcy, Öberg worked with four journeymen (one was his son) and three apprentices.[41]

In the application for bankruptcy Öberg mainly blamed his brother-in-law, the previously discussed book printer Anders Gustaf Hellstén. As mentioned, Hellstén had worked with Per Adam Wallmark until December 1834, when the latter took full control of the task of printing *Stats-Tidningen*. Hellstén tried to stay in business during the latter half of the 1830s, but was eventually forced to apply for bankruptcy in December 1842. The number of creditors in his case was 58, a very high number. The total debt in his bankruptcy was 30,400 rixdollars, the second highest figure among all book printers who went bankrupt between 1780 and 1846.[42]

It was Öberg's involvement in Hellstén's business that forced him into bankruptcy. The most important sources of credit often came from relatives or other people in the same close social or cultural circles. Fredrik Gustaf Engelbrecht, the trustee in Hellstén's bankruptcy, was for example also the main creditor in Öberg's case. His claim of 4,800 rixdollars amounted to 28 per cent of the total claims. The bankruptcies of Hellstén and Öberg offer good examples of the often very complicated and interconnected economic relations between various actors in the book printing trade. When compared to Ortman, who lacked creditors within the book industry when he became bankrupt in 1827, these two entrepreneurs exhibited close economic ties to publishing houses, book shops, book binders and other book printers.

The major asset in Öberg's estate was a large stock of books and other printed materials. The list contained 15,605 unsold articles with a total value of 7,490 rixdollars, whereas the actual print shop had a value of 4,500 rixdollars. The large stock in the estate indicates problems with marketing and distribution. The production in the print shop seems to have worked normally during the years leading up to the bankruptcy.[43]

Conclusions

The introduction of the mechanized machine printing press in Stockholm in 1833 was a very important event in the history of book printing in the city. At the same time, hand-operated printing presses lived on and were used well into the late-nineteenth century. The production of high-quality books, fashioned in the French style that was especially in vogue during the reign of King Gustav III, was only slowly transferred to the industrialized print shops in the 1800s. Mechanized machine printing was in the beginning more important for the production of low-quality books and newspapers in large quantities.

It is also clear that the removal of state-controls of the book printing industry in 1810 had an extremely important effect. After that, an increasing number of book printers in the city had to compete on market terms. Many small newspapers in this period came and went quickly. High-end publications like the first fashion magazine in Sweden – *Konst och Nyhets Magasin för Medborgare af alla Klasser* – produced by the engraver Fredrik Boye and the master printer Carl Deleen, on the other hand successfully managed to find a readership and became an important part of a new landscape of printed products on the Swedish market.

The 1830s was a turbulent era, with many bankruptcies, for the book printers in the city. This was not only caused by the introduction of a new and expensive technology but also happened for several other reasons. The decade saw fierce competition between liberal and conservative newspapers, the establishment of a few large print shops, as well as the demise of many smaller enterprises. The entrepreneurs behind the new, increasingly industrialized print shops used new methods in type foundries, printing methods and the distribution of books and newspapers. The development in Sweden followed an international trend, with similar patterns in other countries in Europe, and in the United States.

It is interesting to note that the Swedish state itself was forced to compete on the market for printed products, which happened during the print wars in the 1830s. The newspaper *Stats-Tidningen* was transformed into the leading conservative paper of the government, but in the end lost out to the entrepreneur Lars Johan Hierta and the liberal paper *Aftonbladet*.

The book printers in Stockholm were a small but very important group of people, who could influence the spread of information in the nation. For this group of individuals, the 1830s became either a decade of failure or of success. Traditional book printers like Henric Fougt the younger and Ludvig Öberg faced difficulties and were forced out of business. Carl Deleen, who belonged to the older generation, managed to thrive for a while, due to the high quality of his products. The future did not belong to him, however. New entrepreneurs, like Lars Johan Hierta and the brothers Carl and Adolf Norstedt, were different and better suited for the future. They explored new ways to market their products and used the latest technology. Slowly, as the nineteenth century evolved, they became the most successful manufacturers of printed material in the country.

Notes

1 Nordin (1881), pp. 259–260; Mosley (2013), p. 145; Weedon (2013), pp. 160–161.
2 Rimm (2009), pp. 29–51.
3 von Wachenfeldt (2015), pp. 164–188; Hayen (2007), pp. 125–126. See also Berg (2005), pp. 326–331; Fontaine (2014), pp. 111–114.
4 Bring and Kulling (1943), pp. 27–31.
5 Klemming and Nordin (1883), p. 238.
6 SSA, Stockholms magistrat och rådhusrätt, F6a, Konkursakter, 142(1/2)/1783.
7 SSA, Rådhusrättens 1:a avdelning (Justitiekollegium, Förmyndarkammaren), FIA, Bouppteckningar, vol. 257, 1779:V:537.
8 Klemming and Nordin (1883), p. 238.
9 SSA, Katarina kyrkoarkiv, FI:5, Död- och begravningsböcker, 1783-02-19.
10 Klemming and Nordin (1883), p. 246. Johan Christofer Holmberg continued as a book printer until 1803.
11 Rimm (2009), III: pp. 5–51.
12 Rimm (2009), II: pp. 4–44.
13 SSA, Rådhusrättens 1:a avdelning (Justitiekollegium, Förmyndarkammaren), FIA, Bouppteckningar, vol. 272, 1783:II:162.
14 Rimm (2009), III: pp. 28. The original letter in RA, Kollegiers m fl, landshövdingars, hovrätters och konsistoriers skrivelser till Kungl Maj:t, Skrivelser från kanslikollegiet med efterföljare, Skrivelser från kanslikollegiet, serie 1, vol. 92 (1782). Translated into English by the author.
15 At the time, she valued the print shop, type foundry and publishing house at 30,000 rixdollars. RA, Ericsbergsarkivet, Svenska autografer, alfabetiskt A-Ö, vol. 142, pp. 75–79, fr. Elsa Momma (Fougt), 25 July 1785.
16 Rimm (2009), III: 31. The original letter in RA, Ericsbergsarkivet, Svenska autografer, alfabetiskt A-Ö, vol. 142, p. 86, Elsa Momma (Fougt). Translation into English by the author.
17 Bring & Kulling (1943), p. 106.
18 Axberger (2016), p. 248.
19 Klemming and Nordin (1883), pp. 393–458.
20 Klemming and Nordin (1883), p. 611.
21 SSA, Stockholms magistrat och rådhusrätt, F6a, Konkursakter, 173/1829. Bring (1934), p. 41. The brother was Carl Anton Ortman, and the wife was Charlotta Carolina Blomqvist.
22 Klemming and Nordin (1883), pp. 418–419.
23 *Journal för Konster, Moder och Seder*, Nr 1, Stockholm 1815.
24 *Konst och Nyhets Magasin för Medborgare af alla Klasser, Första årgången*, Stockholm 1818–1819.
25 The name of the magazine was changed in 1823 from *Konst och Nyhets Magasin för Medborgare af alla Klasser* to *Magasin för Konst, Nyheter och Moder. En månadsskrift*. It continued until 1844.
26 One example is a letter from the editors dated 1 July 1820 in *Konst och Nyhets Magasin för Medborgare af alla Klasser, Tredje årgången*, Stockholm 1820–1821.
27 Wengström (1925).
28 Möller (1945).
29 Nordin (1881), p. 264.
30 Klemming and Nordin (1883), pp. 513–515.
31 SSA, Rådhusrättens 1:a avdelning (Justitiekollegium, Förmyndarkammaren), Bouppteckningar, vol. 417, 1820:IV:286.
32 Klemming and Nordin (1883), pp. 394–395.
33 Weedon (2013), p. 155.
34 SSA, Stockholms magistrat och rådhusrätt, C5a, Konkursdiarier, vol. 12, no. 597. The case file for the bankruptcy is lost.

35 Klemming and Nordin (1883), pp. 442–447; Nordmark (2001), pp. 34–55; Göransson (1937), pp. 296–312.
36 Heurlin (1906), part II, p. 693.
37 Nordmark (2001), p. 35; Appel and Skovgaard-Petersen (2013), pp. 400–401.
38 Weedon (2013), p. 158.
39 Nordmark (2001), pp. 18–125; See also Ledbetter (2007), p. 7.
40 The Regina Database at the Royal Library in Stockholm, works printed by Elmén & Granberg, 1830–1844.
41 SSA, Överståthållarämbetet för uppbördsärenden, Kamrerarexpeditionen, G1BA, Mantalslängder, 1843 (Staden inre), no. 1235. The document is dated 17 November 1842.
42 The case file for the bankruptcy is lost. The highest debt among all book printers between 1780 and 1850, 38,479 rixdollars was recorded in the bankruptcy of Henric Fougt in 1834.
43 SSA, Stockholms magistrat och rådhusrätt, F6a, Konkursakter, 103/1844.

9 Cabinetmakers and chair makers in Stockholm, 1730–1850. Production, market and economy in a regulated economy

Göran Ulväng

Introduction

In a report to the Swedish Diet on the state of the export of manufactured and handicraft goods, the government trade official Johan Westerman Liljencrantz in 1770 noted that:

> For the last couple of years it has been generally esteemed to live comfortably and be provided with beautiful furniture both in the kingdom and in particular in Stockholm, which has encouraged carpenters, cabinet makers and artists to compete to manufacture such furniture for the public, to some extent with a considerable degree of perfection.[1]

Furniture was one of the most important consumption goods during the eighteenth and nineteenth centuries. In both the aristocracy and the middle classes, comfortable homes for pleasure and socializing were seen as an expression of sophisticated taste.[2] In Sweden, during this period, increased prosperity, improved technology and access to global markets meant that existing houses could be expanded with more rooms, fitted with bigger windows, modernized with tiled stoves, as well as crucially, furnished with more comfortable furniture.[3]

Earlier research has to a large extent focused narrowly on the furniture and especially on the more exclusive pieces produced by the most successful furniture makers for the upper classes. In this contribution I will instead widen the focus by discussing furniture makers in a broader social and economic context. The main source is bankruptcy documents, which provide a range of little used information about the finances, social networks and production of furniture makers.

Three sets of questions will be addressed in the following. The first question deals with the basic framework of operation. How was the production of furniture organized in Stockholm between 1730 and 1850; how many artisans were active in the field; how large were the workshops and what kind of furniture was produced? The second question deals with the distribution of furniture. How did the artisans organize the selling of their goods on a regulated market, and how did they reach their customers? The final question deals with the economic situation and how it developed as a result of political and societal changes and shifts.

The deregulation of a regulated economy

During the eighteenth century, Sweden became increasingly integrated in global trade. At the same time the domestic economy grew stronger, following an agrarian transformation and a government that subsidized manufactures and started to deregulate the economy.

This, in turn, increased the demand for consumption goods. The development was seen in all social groups both in towns and in the countryside.[4] As in many other European countries real wages were more or less static until the first half of the nineteenth century.[5] For the land-owning classes, the nobility and freehold peasants, the agrarian transformation caused a growth in their material standard from 1750 onwards. In manors, vicarages and farmsteads the amount of furniture doubled between 1740 and 1860.[6]

The period was characterized by intensified house building, which meant a growing demand for contractors and artisans. Some of the central parts of Stockholm were renewed, partly as a result of devastating city fires, but also to meet the demand for modern, comfortable dwellings.[7] Increased revenues from farming at the same time led to the expansion and improvement of manor houses in the countryside. In the counties surrounding Stockholm around 50 per cent of the manors were provided with new main buildings between 1750 and 1800.[8]

For the landless, the growth of material wealth was much slower. Despite this they still managed to acquire some household goods that were beyond sheer necessity.[9] For day-labourers in agriculture, real wages declined from the 1820s to the 1840s only to pick up after industrialisation took off in the second half of the nineteenth century with a resulting increase in the demand for labour as well as higher levels of consumption.[10] If we compare the economic situation in Sweden with other countries from the 1750s onwards, it seems that, while England was way ahead, economic development and living standards in Sweden were similar to France, except for the nobility, which never reached the same levels of wealth.[11]

A range of regulations that affected Swedish society were overturned during the period. In 1775 free grain trade was introduced and in 1789 most of the tax-exempted land held by the nobility became free to sell to anyone, which gave birth to a modern land market. The breakthrough for a more deregulated and industrialized society happened from the 1810s onwards. In 1809 Sweden lost Finland to Russia. That same year, a new, more liberal constitution was introduced through which the parliament was given more power. In the following decades new legislation was introduced in most areas of the economy, including the right to find savings banks, business banks and joint-stock companies and the abolition of town customs, guild systems and passport requirements.

Manufacturing and distributing furniture – an international outlook

The conditions for making and distributing furniture changed dramatically in the capitals of Western Europe in the eighteenth and nineteenth centuries. In London, the increased demand together with a deregulation of the market led to an abolition of guilds. Carpenters, cabinetmakers, chair makers and upholsterers became entrepreneurs and moved away from specialized production to running workshops that produced complete sets of furniture. The most successful were upholsterers who increasingly controlled the whole manufacturing process and worked closely with architects when designing complete interiors.[12]

The production became more efficient through standardization and the use of templates, and by streamlining the process with highly specialized workers.[13] Furniture was often manufactured in standard models, which could be embellished with different additions, all priced beforehand.[14]

The number of employees increased. The leading manufacturers in London employed 40–50 workers in the late eighteenth century and up to 350 by the mid-1800s.[15] The production of furniture in the British capital was mostly non-mechanized until the mid-nineteenth century. With no rivers to power saw mills, almost all sawing was made by hand. Boards and timber were purchased from the continent or Scandinavia.[16] Steam powered saw mills became more common in the beginning of the nineteenth century, while steam power, planers and other more advanced technologies only became common in carpenter workshops from the mid-1800s.[17] The shops adjacent to the workshops were often furnished as palaces with show rooms and display windows.[18]

In Paris guilds were abolished during the revolution and both manufacturing, retailing and distribution developed in a similar way as in London. The workshops were owned by merchants or craftsmen, many of them upholsterers, who sold both newly manufactured furniture as well as second-hand objects. The shops were furnished as private homes where buyers could walk around.[19]

Towns in general during this period increasingly became urban centres dominated by trade and craft, with prosperous middle and upper classes demanding luxury goods. Shops got fixed addresses and storefronts and special shopping streets emerged where shops were mixed with cafes and pubs. Shopping became a pleasure.[20]

Manufacturing furniture in Stockholm 1730–1850

With a population of more than 60,000 inhabitants, Stockholm was by far the largest city in Sweden in the mid-eighteenth century. It was the primary administrative centre and also one of the most important commercial centres. The city was the centre for foreign trade with a substantial burgher population consisting of both merchants and artisans. Many in the latter group were specialized in luxury production for the elites in Stockholm and the wider kingdom. About 12 per cent of the city's inhabitants were noble or non-noble persons of rank compared to about 3 per cent in the country as a whole.[21] Artisans and merchants were the largest group by numbers. One-third of all males in Stockholm were shoemakers, tailors or goldsmiths, while grocers and drapers dominated among the retailers. The latter group included a number of minor trades, such as fruitmongers and cloth brokers.[22]

Stockholm was not only a large city with a diverse business and trades community but also a city with huge problems. Wars with Russia and Denmark in the beginning of the eighteenth century and an associated pestilence affected the city hard. While it recovered in the following decades, it again fell into a long period of economic and social stagnation which lasted for nearly a hundred years. The rest of the country on the contrary saw a steady growth in agriculture, trade and industry.[23]

Even though experiencing sluggish growth, Stockholm still was the most important centre for the production and import of both general and especially luxurious goods. Some researchers have pointed out that the number of dealers and craftsmen actually remained the same or even increased.[24] Boëthius showed that the number of artisans nearly doubled from 1,100 to 2,000 between 1740 and 1815; the number of dealers rose from 530 to 750 and that 10 manufacturers in 1740 had increased to 320 in 1815.[25]

The market for furniture production in Stockholm was highly regulated until 1846. The guilds for cabinetmakers, chair makers, turners and similar trades strove for monopolies to secure the income of their members and maintain the quality of their work. To avoid competition the workshops and shops had strictly limited opening hours, and advertising and price currents were more or less forbidden, as it was perceived as something utterly unfair.[26] The only carpenters who were allowed to work outside of a guild in Sweden were those who lived in the countryside, who were soldiers, were working in a noble family or employed at the Royal Castle, but they were extremely few in Stockholm.[27]

In 1722 the government introduced hallmark courts (*hallrätter*), municipal supervisory bodies with the sole right to authorize, oversee the production in and control the quality of the goods in urban manufactures, with the aim of widening the production without competing with the traditional guilds.[28] The woodworking guilds opposed the hallmark courts, with the argument that they encouraged illegal competition and threatened the economy and the quality of the products by allowing non-guild furniture makers to establish in Stockholm.[29]

The average workshop was a small-scale operation. Furniture makers in general had limited revenues. With high costs for material and labour most of them only made furniture on commission and couldn't afford to build up a stock. An average furniture maker in eighteenth-century Stockholm had two to five employees, including journeymen and apprentices.[30]

Furniture makers – number and production

The number of cabinetmakers and chair makers in Stockholm has until recently been quite uncertain. Only after collecting and combining biographical information from a range of sources, have I been able to get a statistical picture of the number of woodworking artisans as well as the size of their workshops.[31]

When processing this data, several interesting features have been detected. Starting with the master cabinetmakers, their numbers grew from 40 in 1730 to 90 in 1780, thereafter declined until 1810 but grew again to 90 in 1840 (see Figure 9.1). As already mentioned, most workshops were small with on average only two to five employees.[32] While there are no records of the number of journeymen employed, the number of apprentices generally correlates with the number of cabinetmakers. The exceptions include the period up until 1760, the year 1800 and the period after 1825, three periods when the number of apprentices grew more than the number of masters, in effect an indication that the workshops grew in size.

When considering how the development corresponds quite well with the growth and decline of the gross domestic product (GDP), a general conclusion would be that furniture was a price sensitive item. The Swedish GDP increased between 1750 and 1780, decreased from 1780 to 1810 (save for a couple of years around the year 1800) due to crop failures and lower real wages and rose again after 1810.[33]

The only exception to this agreement is a decrease in the number of apprentices already from the 1760s. This can be explained by the financial crisis that hit Stockholm in the 1760s, which forced masters to reduce production, and in some cases subsequently also saw them closing down their workshops. The skilled cabinetmaker Lorentz Nordin gave up his business in 1764 "due to the costly times with journeymen wanting to be paid double".[34]

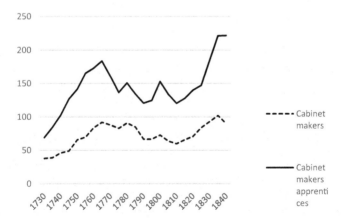

Figure 9.1. The number of cabinetmakers and their apprentices in Stockholm, 1730–1845

Source: Sylvén (1996).

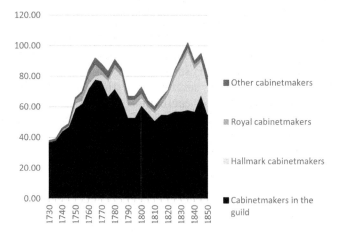

Figure 9.2. The number of cabinetmakers affiliated to the guild or to other institutions in Stockholm, 1730–1850

Source: Sylvén (1996).

The fluctuations in the number of active cabinetmakers were primarily
due to shifting numbers authorized by the hallmark court. The Carpentry
Guild on the contrary tried to restrict the number of members and as a
result remained relatively constant during the period, with around 60 members. The only exception was a few decades from the 1750s onwards when an
increased demand led to expanding production and a growth of the membership to 80 members (see Figure 9.2).

The guild became more restrictive in appointing masters after 1770, and
the numbers decreased as a result. Interestingly the total number of cabinetmakers still remained almost the same as the guild-appointed masters were
replaced by non-guild cabinetmakers.

The number of chair makers developed in a different way and was generally less marked by fluctuations. Their numbers grew until the financial
crisis in the 1760s, after which they entered a long period of gradual decline,
only to increase again after 1820 (see Figure 9.3). When compared to the
cabinetmakers, the chair maker workshops were larger with between eight
to ten employees.

I would propose three reasons why chair makers were less affected by the
economic fluctuations than the cabinetmakers. First, their guild was highly
restrictive when appointing members, which led to less competition and less
vulnerability. Second, they mostly used native types of wood, which made
them less dependent on expensive imports than cabinetmakers who had to
secure exotic woods and buy expensive metal fixtures. Third, chair makers
benefitted from a strong and solid demand for their products, as a direct result of the prominence of chairs and sofas as fashionable interior decoration
elements during the period.

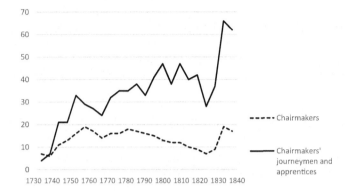

Figure 9.3. The number of chair makers and their journeymen and apprentices in Stockholm, 1730–1845

Source: Sylvén (2003).

Much like in Britain, few major technological innovations affected furniture making during the period. When considering that the number of masters, journeymen and apprentices involved in cabinet and chair making increased from 122 in 1730 to over 400 in 1840, this would suggest that the production simply increased three times. There are reasons to believe that the increase was stronger than that, however, as a result of minor technological changes.[35] It is well known that chair makers in Stockholm started to use decorative plaster on their chairs from the early 1800s instead of carving the ornaments. The technique not only made possible a greater variation in the ornamentation, but also saved time.[36] This use of plaster likely facilitated the production of larger series of chairs which, in turn, supported the growth of the workshops. In another example, the renowned chair maker Efraim Ståhl had his workshop organized in a continental way with separate rooms for carpenters, guilders, sculptors and upholsterers.[37]

Stockholm cabinet and chair makers were celebrated for their high-quality furniture. Many of the masters and journeymen had trained abroad and closely followed the development in architecture and design, which was then adapted to Swedish tastes and demand. The beautifully veneered cabinets, drawers and tables that were produced in their workshops have been highlighted and described by art historians and museum curators as Stockholm pieces (*Stockholmsarbeten*) to distinguish them from furniture manufactured in smaller towns, which were made from cheaper native woods, and often painted.[38] In my opinion, the distinction between the furniture produced in the capital and provincial works is misleading in many ways. A study of 35 probate inventories from workers, craftsmen and officials in Stockholm in 1780 show that painted furniture were dominant in all homes, including those of wealthy, noble families. Of the over 850 furniture in the inventories, only 32 were veneered, most of them drawers.[39]

Most of the painted furniture must have been manufactured in Stockholm. Some could have been bought in other towns or brought to Stockholm when families moved to the capital. When considering the high costs for transportation, the number of pieces introduced this way was likely low, however. The conclusion must be that most of the furniture makers in Stockholm had a wide range of customers from all social groups, with the exception of a few who were specialized in high-quality pieces for the royalty and for wealthy noble and merchant families.

The broad clientele can be seen in the bankruptcy documents of Erik Neibert, a distinguished Stockholm cabinetmaker. His lists of creditors, many of whom likely also were his customers included traders, craftsmen and civil servants.[40] The idea of an exclusively high-end production in Stockholm can also be challenged when considering that many of the workshops in the capital produced panels, doors, window frames for normal houses and pulpits, altar pieces and benches for churches.[41]

The distribution of furniture

The sale of furniture was for a long time an exclusive right held by the cabinet and chair makers in Stockholm. Sales were largely based on commission orders since the high costs for both materials and the running of the workshops meant that few producers could afford to keep a stock of goods. If dealers were allowed to sell furniture, there was a risk that they would buy goods from non-guild producers, which would undermine the position of the masters.[42]

In 1739, new legislation was introduced, which allowed for the operation of manufactures in all parts of the country. This was seen as a serious threat to cabinet and chair makers, since their monopoly included the sole right to sell furniture outside of the cities, including at rural fairs.[43] The bankruptcies after the cabinetmakers Alexander Thunberg and Erik Neibert, in 1782 and 1785, respectively, show that many of their creditors and likely customers to the artisans in the capital indeed were merchants, craftsmen and ship owners from small towns around Lake Mälaren in the Stockholm hinterland as well as in Finland.

When demand increased in the 1730s and 1740s some of the cabinet and chair makers started to sell their products directly from their workshops. In the 1750s, the guilds also established special shops, where the guild members could sell their products. Furniture and chair makers appointed by the hallmark court had their own shops and as a consequence could sell more freely.[44]

In the 1760s several furniture shops of the same kind as those in London and Paris were established in Stockholm. These shops were not founded by the producers, but instead by merchants who took on a role as middlemen between the producers and the buyers. The new vendors in some cases also engaged in the export of furniture to the Baltic region and Germany.[45] The new shops enabled a higher degree of specialisation essentially by allowing masters to focus on production, instead of marketing. The development becomes clear when studying documents from furniture makers' bankruptcies. After 1785, these documents altogether lack creditors from outside of Stockholm, which suggests that vending was handled in the capital through different channels.[46]

Most of the new furniture shops ceased their operation in the late eighteenth century. This was partly a result of weakened demand at that time. The second reason was that grocers, cloth merchants and ironmongers were given the right to sell certain types of goods, including furniture and mirrors.[47]

This can be confirmed by bankruptcy documents from furniture makers, which shows that almost all of their furniture were sold in shops owned by various mongers or grocers.[48] When the cabinetmaker Erik Neibert filed for bankruptcy in 1782, his furniture was sold in various shops across the city. These included Madame Sundstrand's shop located at Skanstull on the southern island of Södermalm; a cloth stand in a house owned by the actuary Zetterman; a cloth stand in a house owned by Herpell as well as two premises located in the Old Town: Madame Rosling's shop on Österlånggatan and in a cloth stand in a house overlooking Slussen.[49]

In general, trade and manufacturing in Stockholm became more diverse, and old rules were challenged during this period. When the cabinetmaker Johan Petter Berg went bankrupt in 1809, one of his creditors was a spice grocer, who lodged a claim for compensation, not as would perhaps be expected for spices, but for four-tiled stoves that he had sold to, and installed in Berg's home.[50]

A third reason why the furniture shops went out of business was that some of the furniture makers begun to sell from their own shops. Many early nineteenth-century bankruptcy documents reveal that furniture makers kept products in warehouses, sometimes quite valuable items, in a way which suggest that they increasingly handled the sales on their own.[51]

The guilds kept some control over the way in which sales were conducted. Unlike shops in London or Paris a typical shop in Stockholm lacked an entrance from the street or even a storefront, and the makers were not allowed to advertise their prices.[52] Trade cards were not used and newspaper advertisements were more or less non-existent. As late as 1870, when the guild system was long abolished and freedom of trade was a reality, advertising in newspapers was still a rarity.[53]

A final reason why furniture shops were long-term unsuccessful was the competition from the auction market. Auctions were important fixtures in the regulated Stockholm economy and were held four to five times a week. The dominant items offered to the market were textiles, clothes and household utensils, but also furniture.[54]

The economy and network of the furniture makers

The Swedish art historian Marshall Lagerqvist claimed that mid-eighteenth-century furniture makers in Stockholm were successful and wealthy. As shown by the economic historian Ernst Söderlund, most furniture makers in reality were among the poorest artisans in the city. They not only engaged in time-consuming manufacturing with associated high labour costs for journeymen and apprentices but also used expensive materials, often imported from abroad.[55]

Through an analysis of bankruptcy documents, I would argue that it is possible to gain new insights into their economic situation. Between 1768 and 1831 as a whole, 22 furniture makers applied for bankruptcy (see Table 9.1). A first observation would be that relatively few furniture makers actually faced bankruptcy, despite the fact that they arguably were among the poorest artisans. Indeed, the 22 individuals only comprised some 3 per cent of all the active furniture makers during this period. Even when focusing on three periods marked by weak demand in the business between 1780–1790, 1800–1810 and 1820–1830, no more than 5–6 per cent of all furniture makers applied for bankruptcy (see Table 9.2).

It is generally difficult to know if this was a low or high proportion of bankruptcies for a specific professional group, as there are few studies on

Table 9.1 The number of cabinetmakers and chair makers in Stockholm who went bankrupt, 1757–1831

Year of Bankruptcy	Name	Surname	Title	Act number
1768	Lars	Boberg	Cabinetmaker	67/1772
1779	Johan Magnus	Börling	Cabinetmaker	90/1780
1780	Anders	Blom	Chair maker	171/1780
1781	Eric	Neibert	Cabinetmaker	213/1782
1781	Alexander	Thunberg	Chair maker	40/1782
1782	Catharina	Östberg	Chairmaker's widow	57/1784
1783	Johan	Lötström	Cabinetmaker	36/1784
1784	Eric	Neibert	Cabinetmaker	148/1785
1791	Carl Fredr.	Flodin	Chair maker	171/1792
1791	Cath. Elis.	Thunberg	Chairmaker's widow	134/1792
1802	Fredric	Schalin	Cabinetmaker	227/1803
1807	Isaac	Löfgren	Cabinetmaker	66/1808
1807	L. Eric	Engström	Cabinetmaker	100/1808
1807	Niclas	Groth	Cabinetmaker	67/1808
1808	Johan Peter	Berg	Cabinetmaker	71/1809
1819	Magnus	Rosengren	Cabinetmaker	309/1819
1820	Ephraim	Ståhl	Chair maker	353/1822
1820	Johan Melch	Lundberg	Chair maker	189/1822
1825	Jacob David	Jacobsson	Cabinetmaker	93/1826
1825	Johan	Vougt	Cabinetmaker	183/1826
1831	Johan Christian	Wieckman	Chair maker	126/1832

Table 9.2 The number of furniture makers in Stockholm 1760–1840 the number of furniture makers who filed for bankruptcy and the reasons for their decision

Period	Number of furniture makers	Bankruptcies		Main reasons
		Number	Per cent	
1760–1770	103	2	2	-
1770–1780	102	3	3	Beginner (1), the business run by his wife (1)
1780–1790	105	5	5	No demand (3), beginner (1), debts (1), illness (1), drunkenness (1)
1790–1800	82	2	2	No demand (2), illness (1)
1800–1810	80	5	6	Losses in business (3), beginner (1), illness (1), debts (1), accident (1)
1810–1820	74	1	1	Accident (1)
1820–1830	85	5	6	No demand (2), beginner (1), being a Jew (1), broken (1)
1830–1840	110	1	1	No demand (1), beginner (1)

Source: SSA, Stockholms magistrat och rådshusrätt, F6a, Konkursakter, huvudserie.

bankruptcies in other professions. We also for example lack information about the number of furniture makers who quit their businesses because of financial difficulties, but avoided bankruptcy. Bankruptcy was probably socially stigmatizing for most and something that was avoided for the longest time.

The reasons why furniture makers went bankrupt were in most cases due to a lack of demand. Many of them also referred to the fact that they were new to the business and had difficulties to obtain commissions. The cabinetmaker Johan Magnus Börling claimed that the main reason for his bankruptcy was the fee that he had to pay to the guild in order to obtain his privileges. As a result, he was forced to borrow money to be able to purchase tools and materials, but with limited demand at that time, was unable to make the repayments.[56]

It is well known from previous research that it was difficult to establish a workshop in Stockholm. Often, it took some 20 years for a journeyman to obtain rights to practice as a master. This led many to open businesses in smaller towns, where the competition was lower, or to venture abroad, some of them to London.[57] Illness and accidents were also common reasons for bankruptcies, two factors which struck even harder in times of a general economic downturn when demand was lower and margins narrower.

The bankruptcy records are unique, since they contain information about debts which cannot be obtained elsewhere. Whereas probate inventories in most cases contain information about the creditors and the amounts owed, the bankruptcy documents go further by providing details of the nature of the debts, making it possible for example to see if it was in the form of cash, goods, food, rent or anything else. Obviously, this was not a debt structure fully representative of normal life. Cabinet and chair makers in most cases were in highly precarious economic situations when they filed for bankruptcy, with many debts and decimated assets. The inventories show how they even had resorted to pawning finished pieces of furniture to the creditors or sold off both household utensils and clothes.[58]

When combining information from various documents gathered during the bankruptcy proceedings, it becomes possible to follow changes in the number of creditors, the size of the debts and the assets of the debtors (see Table 9.3).

In the following analysis, in order to gain a complete picture of the economic situation, I have included the total number of creditors, reported at various stages during the process. This include (1) those who were reported as creditors by the debtor at the initiation of the proceedings, but never turned up to guard their claims at the day of proclamation, when the case got underway, and (2) those who turned up at the day of proclamation, and only at this point were registered by the court as creditors. The creditors in the latter group in most cases could show legally binding documents that supported their claims, such as bonds, signed by both the lender and the borrower, and witnessed by reliable persons.

Two conclusions can be drawn. First, the number of creditors in the latter group increased from 5 to 25, calculated as median value. Second, this happened despite a much weaker growth in the number of creditors reported

Table 9.3 The number of creditors, the size of the debts and the debtors' assets in mean and median values, 1770–1830, in fixed prices (Swedish Consumer Price Index (CPI), 1914=100)

Period	The number of creditors, according to the debtor.	The number of creditors according to the Town Hall Court, and the size of the debts.		The assets of the debtor.
	Number	Number	Sum	Sum
Mean				
1770–1790	17	9	8129	3758
1790–1810	20	16	5583	7298
1810–1830	15	31	16442	9085
Median				
1770–1790	15	5	5516	685
1790–1810	17	12	1927	927
1810–1830	16	25	4392	1752

Source: SSA, Stockholms magistrat och rådshusrätt, F6a, Konkursakter, huvudserie.

by the debtor. In the period 1770–1790 only five out of the 15 creditors reported by the debtor in the end actually claimed their loans. In the period 1810–1830 no less than 25 creditors turned up to claim their loans, a significantly higher number than the 16 that the debtors originally reported. The reason should be sought in the improvement of and greater trust in the bankruptcy institution, which, in turn, made creditors more likely to use it to secure outstanding claims. The process could be interpreted as a transformation from an informal family/household-based economy into a more modern institutionalized economy.

It also becomes apparent that the creditors during the latter period were likely to get more of their money back. The debts were in fact considerably lower in relation to the debtors' assets in the first decades of the nineteenth century than 50 years earlier. In the period 1770–1790, the median assets for furniture makers were 685 rixdollars, while the debts were 5,516 rixdollars or in short nearly eight times higher. In 1810–1830, the corresponding relation was 1,752–4,392 rixdollars. One possible reason for this could be the growing number of creditors, which increased the pressure on the debtors and forced them to file for bankruptcy earlier than before (see Table 9.3).

Interesting changes over time can also be seen by analysing the types of debts (cash, goods, food, rents) that were reported in the bankruptcy documents as well as the social categories of the people to which the furniture makers were indebted.

The 22 furniture makers who filed for bankruptcy during the period had altogether 553 creditors. I have classified them in six groups based on titles and social categories: employees (farmhands, apprentices, dayworkers), artisans (masters, carpenters, turners, mirror makers), dealers (grocers, mongers, merchants), officials (officers, customs officers, clerks), institutions (the city, the state) and others.

I have chosen to make separate studies of the cabinet and the chair makers, since they had very different kind of economies, where the former were much more dependent on the import of exclusive wood and metals and had higher costs for manufacturing. I am also interested if there were any differences over time, but the small number of debtors means that the results have to be interpreted carefully (see Tables 9.4 and 9.5).

A first observation is that the cabinetmakers were indebted to individuals from many different social groups, including employees, craftsmen, dealers, officials and institutions. This pattern is different from what we know about financial networks among farmers, manor owners and merchants who usually mainly borrowed money from people within their social group.[59] Cabinetmakers were on the contrary involved in several networks as both buyers of a wide range types of materials and goods, and as sellers, to a wide range of customers in all social groups.

The dominant kind of debt among the cabinetmakers was cash debt, which made up 83 per cent of the total debt in the late eighteenth century

Table 9.4 The share of the debts, in relation to the creditors' belongings, in bankruptcies among cabinetmakers 1770–1800 and 1800–1830, in fixed prices (Swedish Consumer Price Index (CPI), 1914=100)

Period	Creditors	Total	Debts		Salaries and materials		Rent or house maintenance		Food		Consumption	
		% of the sum of debts	% of the sum of debts	Mean value debts	% of the sum of debts	Mean value debts	% of the sum of Debts	Mean value debts	% of the sum of debts	Mean value debts	% of the sum of debts	Mean value debts
1770–1800 4 cabinetmakers, 57 creditors	Employees	5	4	869	1	75			0	76	2	83
	Artisans	43	41	628			1	158	2	200	9	428
	Dealers	30	18	524			1	172				
	Officials	14	13	614								
	Institutions	8	7	214								
	Sum	**100**	**83**		**1**		**2**		**2**		**11**	
1800–1830 7 cabinetmakers, 116 creditors	Employees	0										
	Artisans	40	19	252	3	71	4	218	11	73	3	60
	Dealers	20	13	217	5	141	1	155	0	111	1	46
	Officials	31	30	502			1	10				
	Institutions	9	8	641								
	Sum	**100**	**70**		**8**		**6**		**11**		**4**	

and 70 per cent in the early nineteenth century. The creditors were, with few exceptions, craftsmen, such as carpenters, brewers, watchmakers and restaurateurs, and dealers, including victual dealers, spice grocers and wholesalers. These two groups also dominate, among those who had claims based on lack of payment for delivered food and materials. These debts, what in modern accounting would be referred to as accounts payable, or in simpler terms delivery debts, constituted the rest of the debts besides the cash debts, at 17 and 30 per cent for the different time periods, respectively.[60]

The main part of the delivery debts related to consumption goods (textiles, clothes, furniture), food (beer, liquor, fish, bread) as well as rents for dwellings and workshops, and the upkeep or addition to dwellings (installation of tiled stoves, repairs of windows, painting jobs).[61] A likely reason why a higher share of delivery debts can be observed in the latter period than in the earlier period would be that the creditors became more interested in monitoring their claims, as they were based on more secure bonds, and hence more likely to be successfully repaid.

The bankruptcies among the chair makers seem to follow a different pattern. Most of the late eighteenth-century chair makers had relatively few debts when compared to the cabinetmakers. Chair makers had almost no debts to dealers and very few to other craftsmen; instead we find officials and institutions among their creditors. The chair makers were generally less dependent on a wide range of dealers and manufacturers than the cabinetmakers, since they mainly produced chairs made from native timbers. There was in other words no need for the type of exotic imported timber, special fittings or limestone tops, which the cabinetmakers had to purchase, from various sources. We also have to bear in mind that the late eighteenth-century chair makers to a larger extent made their products on commission, with a more limited exposure to dealers. This is probably the reason why there are so many officials among the creditors; the chair makers borrowed from their clients (see Table 9.5).

The bankruptcy documents show that the chair makers who filed for bankruptcy in the early eighteenth century had widened their production. By this point they not only manufactured chairs but also other kinds of furniture and had also started to produce for a general market instead of just working on commission. The larger market exposure is obvious in the documents. The chair makers now had debts to craftsmen and dealers, just like the cabinetmakers, and a large share of the debts were delivery debts. Since many of the chair makers ran bigger workshop, the bankruptcies affected a large number of creditors as well as employees. The bankruptcy of the widow of Efraim Ståhl, the most famous of the early nineteenth-century chair makers, reveals how delicate it was to run a workshop of this size in Stockholm. Despite his many clients, among them the Royal Court, members of the nobility and institutions, he was constantly in debt and dependent on the kindness of his subcontractors.

Table 9.5 The share of the debts, in relation to the creditors' belongings, in bankruptcies among chair makers 1770–1800 and 1800–1830, in fixed prices (Swedish Consumer Price Index (CPI), 1914=100)

Period	Creditors	Total	Debts		Salaries + Materials		Rent or House Maintenance		Food		Consumption	
		% of the sum of debts	% of the sum of debts	Mean value debts	% of the sum of debts	Mean value debts	% of the sum of debts	Mean value debts	% of the sum of debts	Mean value debts	% of the sum of debts	Mean value debts
1770–1800 3 chair makers, 15 creditors	Employees	4	0	41	4	572						
	Artisans	6	5	345					1	90		
	Dealers	0										
	Officials	67	67	2386								
	Institutions	23	23	1048								
	Sum	**100**	**95**		**4**				**1**			
1800–1830 3 chair makers, 137 creditors	Employees	2	0	69	1	172	0		4	303	7	731
	Artisans	50	4	269	33	1737	2	368	1	284	5	464
	Dealers	26	16	1853	4	368	0	319			0	185
	Officials	10	9	737	1	256		132				
	Institutions	12	11	3887								
	Sum	**100**	**40**		39		2		5		12	

Conclusions

The aim of this chapter was to highlight the conditions for furniture makers in Stockholm between 1730 and 1850 with a focus on their production, market and economy. By using bankruptcy documents, I have been able to give a new and more nuanced picture of their financial situation during the period.

Furniture was one of the most important and most demanded goods during the period of investigation. Despite a stagnant development of the Stockholm economy in general, the number of furniture makers, journeymen and apprentices increased from 100 to more than 400 during this period. The growth was a consequence of legislative changes, introduced to meet the increased demand. While the number of furniture makers appointed by the guilds remained stayed the same, furniture makers approved by the hallmark court, an institution established to improve the manufacturing of goods in general, grew in numbers. Above all, it was the number of cabinetmakers that grew in number, while the chair makers remained relatively few. The chair makers' guild seems to have had a stronger control over their members to avoid over-establishment.

Most workshops remained small, with only three to four employees, and worked mainly on commission, producing not only furniture but also other furnishings for a wide range of customers from all social groups. Most of the furniture that was manufactured was made of indigenous wood that was painted. Only a few per cent of the furniture were of a more exclusive kind, made of imported wood or veneered or varnished. Especially the furniture makers had high costs for importing materials and a time-consuming, labour intensive, manufacturing process, which meant low-financial returns. This was further aggravated by guilds that controlled sales to avoid competition and prohibited advertising.

The way, in which furniture was manufactured and sold in Stockholm was dated when compared to the deregulated situation in London and Paris, where workshops could have hundreds of employees; include showrooms and where the use of trade cards and advertisement was normal. In Stockholm, auctions became a way to negotiate this imperfect market. By the late eighteenth century, auctions as a result were extremely popular, as places where furniture and other goods could be sold and purchased more freely. The furniture makers seem to have a greater interest in both marketing and selling their products compared to the chair makers. The furniture makers travelled to markets around the country and when traders in Stockholm established special furniture stores in the 1760s, it was the furniture makers who sold their furniture there. This was probably due to the great competition that existed between the furniture makers and thus a great need to be able to sell the sometimes very exclusive furniture. The chair makers were still mostly working on commission by the end of the eighteenth century.

The establishment of the hallmark court was a first step towards a more deregulated market. In the second half of the eighteenth century, mongers and grocers were given the right to sell a wider range of goods, and they soon became important distributors to the furniture makers. Some of the furniture makers also opened up shops in the early 1800s to be able to sell directly from their stocks. It is obvious that the market had become more mature, with a more stable demand and possibly higher return.

Even though most furniture makers were quite poor, very few went bankrupt. Most of the furniture makers who went bankrupt did so when times were difficult and the competition grew. Those who suffered the most were the newcomers, who recently had established a workshop.

Most cabinetmakers formed part of networks that consisted of subcontractors, colleagues, grocers, mongers, sellers and buyers, which branched out across the local society. This also made them relatively strong and reliable. Interestingly, late eighteenth-century chair makers did not have these kinds of wide networks, since they to a lesser degree used imported materials and fittings, but instead used their customers as creditors. This independence seems to have made them more vulnerable.

Notes

1 The translation from Swedish by the author. Lagerquist (1981), p. 37.
2 Hellman (1999); Edwards (2005); Berg (2005), pp. 250–255; Stobart and Prytz (2018).
3 Ulväng (2004).
4 Ahlberger (1996); Müller et al. (2010).
5 Trentmann (2004), p. 381.
6 Ulväng (2004), pp. 147, 179, 215.
7 Lagerquist (1981), p. 37; Danielsson (1998).
8 Ulväng (2011).
9 Hallén (2009), pp. 161–170.
10 Magnusson (1997), p. 309.
11 Hallén (2009).
12 Edwards (1996), pp. 20, 35, 50; Kirkham (1988), pp. 57–65; Coquery (2004), p. 78.
13 Edwards (1996), p. 25.
14 Edwards (1996), pp. 27–29.
15 Kirkham (1988), p. 72.
16 Edwards (1996), p. 81.
17 Kirkham (1988), pp. 109–113.
18 Edwards (1996), p. 52.
19 Coquery (2004), pp. 78, 87.
20 Coquery (2004); Sargentson (1998); Fairchilds (1993); Mui and Mui (1989); Walsh (2014).
21 Carlsson (1949), pp. 35–40.
22 Eggeby and Nyberg (2002), p. 214.
23 Generally on the stagnation, see Eggeby and Nyberg (2002), p. 187f.
24 Wottle (2000).
25 Boëthius (1943), pp. 32, 441. A parallel to Stockholm is Antwerp, which during a long stagnation period saw the establishment of an increasing number of artisans, from 2,000 serving 100,000 inhabitants in the mid-sixteenth century to 3,000 serving 50,000 people in the late eighteenth century, as a result of the consumer revolution. Blondé and van Damme (2010).
26 Söderlund (1943).
27 Söderlund (1943), pp. 293–300.
28 This legislation was supplemented in 1739 with the goal to expand the number of manufactures. Söderlund (1943), p. 298.

29 Boëthius (1943), pp. 96–100.
30 Söderlund (1943), p. 234.
31 Ulväng (2017), pp. 118–122.
32 Söderlund (1943), p. 234.
33 Schön and Krantz (2012).
34 "til följe av tidernas dyrhet at öka gesällernas arbetslöner til dubbelt". Lagerquist (1981), p. 37.
35 The knife makers in Stockholm for example improved their production by establishing more effective ways of organizing the production according to English models, partly by using watermills. Jansson (2017).
36 Sjöberg (1991), p. 10.
37 SSA, Stockholms magistrat och rådhusrätt, F6a, Konkursakter, huvudserie, 353/1822.
38 Holkers (2008); Nyström (2008), p. 16; Falck (2008a), p. 84; Martinius (2008), p. 155.
39 Ulväng (2017).
40 SSA, Stockholms magistrat och rådhusrätt, F6a, Konkursakter, huvudserie, 213/1782.
41 Hyvönen (1988). Another important and probably profitable product was coffins, as can be seen on business signs from the period. A sign from 1796 from Carl Henrik Blom's workshop in Stockholm, Nordiska museet, inv. nr. NM.0080338, and a sign from 1790 from Christopher Bonsdorff in Finland, Kuopio historiska museum, inv. nr. 1205, in Hyvönen (1988), p. 94.
42 Söderlund (1943), pp. 224–230.
43 Söderlund (1943), pp. 222, 224; Lagerquist (1981), pp. 76, 98–100.
44 Lagerquist (1981), p. 71f.
45 Lagerquist (1981), pp. 102–105, 142.
46 At the same time the production of furniture increased in the small towns in the region. Söderlund (1943), p. 268.
47 Sylvén (1996); Lagerquist (1981), p. 76; Wottle (2000). Murhem (2016) shows that most of the advertisements for selling furniture were published by mongers and grocers, beside the Stockholm Auction House.
48 Ulväng (2017).
49 SSA, Stockholms magistrat och rådhusrätt, F6a, Konkursakter, huvudserie, 213/1782.
50 SSA, Stockholms magistrat och rådhusrätt, F6a, Konkursakter, huvudserie, 71/1809.
51 SSA, Stockholms magistrat och rådhusrätt, F6a, Konkursakter, huvudserie, 67/1808; 71/1809; 353/1822.
52 Lagerquist (1981), p. 98.
53 Larsson (2016).
54 Ulväng et al. (2013).
55 Söderlund (1943), p. 207.
56 SSA, Stockholms magistrat och rådhusrätt, F6a, Konkursakter, huvudserie, 90/1780.
57 Söderlund (1943), p. 278f.; Wood (2006).
58 SSA, Stockholms magistrat och rådhusrätt, F6a, Konkursakter, huvudserie, 67/1772; 171/1780; 213/1782; 71/1809; 126/1832.
59 Hasselberg (1998); Ågren (2007); Eriksson (2018).
60 It is possible that some of the cash debts actually were delivery debts at the outset, and only later were transformed into cash debts. It is impossible to trace this in the existing material, however.
61 There was no sharp boundary between the private sphere and the business sphere, since the position and the respectability were determined by the lifestyle and consumption. Hasselberg (1998), p. 276f.

10 Credit relations among painting professionals in Stockholm, 1760–1849

Axel Hagberg and Klas Nyberg

Introduction

Painters, sculptors, gilders and wallpaper painters formed an important group among the artisans who contributed to the production and spread of new luxury in eighteenth-century Stockholm. In Sweden, professional artistic and artisanal painting, including wall painting, sculptural painting, decorative painting and figurative painting, was generally carried out by town-based guild-connected artisans. Wallpaper painters in contrast worked in manufactures, were treated as industrial workers and instead answered to the municipal hallmark courts. When conflict between the two groups arose, with wallpaper painters offering industrially produced wall decorations that encroached on the privileges of the guild-painters who provided decorative oil-painted wall decorations, the latter were generally favoured by the authorities.

Sculptors and gilders were also regulated by the hallmark court. Painters who worked for aristocratic households or special so-called "court painters", on the other hand, were outside of guild control. By the end of the seventeenth century, one third of all painters and apprentices worked outside of the guilds in noble households. Various wandering self-taught amateur painters, who roamed the countryside, painting in a folk art tradition as well as rural furniture painter were neither regulated by guilds nor the hallmark courts. Painters who worked in the porcelain manufactures in Stockholm, to decorate faience, were counted as manufacture workers and answered to the hallmark courts.

Painting and the associated arts in Stockholm were affected by two contradictory trends during the eighteenth and early nineteenth centuries. The sector was favoured by a growing market for art and decoration on a national level, but was at the same time hit by Stockholm's economic stagnation. The market for painters grew when decoration demands increased as burghers as well as the aristocracy decorated new urban homes as well as rural manor houses and estates. The market for depictions and paintings was also growing and the demand for portraits increased when the growing elites adopted more conspicuous lifestyles. Painters working in Stockholm must at the same time have been affected when their home market con-tracted as the capital stagnated financially from the 1760s onwards. As the Swedish capital and as the centre of politics, Stockholm still was a unique place when compared to the rest of the country, with a high concentration of affluent individuals with conspicuous lifestyles who created a demand for painting, painted decorations and portrait painting, as part of a general consumption of luxury.

In the following, the role of painting professionals in Stockholm, including painters and similar professions such as sculptors, gilders and wallpaper makers will be investigated during the period between 1760 and 1849. The investigation is set against a background which takes into account the changing economic situation and shifting credit relations that developed in Stockholm from the second half of the eighteenth century onwards.[1] The study provides a statistical analysis of the larger painting community, but mainly focuses on bankruptcy records, to determine how members of the painting community developed credit relations and their social and eco-nomic networks during a period of gradual modernization of the Swedish economy. In the investigation below, we seek to answer if it is possible to find indications of a modern financial argumentation in the bankruptcy files in-volving painting professionals during the first decades of the nineteenth cen-tury, i.e. in the period after the reform of the Swedish bankruptcy institution between 1767 and 1818. A specific focus is on the use of surety commitments or in other words personal sureties for loans.

The number of painting professionals, their wealth and bankruptcies

Well-preserved tax ledgers offer a good idea of the number of painting professionals and their income levels in Stockholm during the period. The numbers do not show the entire community. According to census registers, the total painting community, here taken to mean painters, sculptors, gild-ers and wallpaper makers, was composed of some 50 individuals in 1740. The number of individuals who paid taxes was lower, however. A number of reasons for this can be presented, including general poverty, inactive businesses and limited work due to old age.

The information from the tax ledgers shows that the income level among the painting community was relatively homogeneous between 1720 and 1770 (see Table 10.1). When excluding the partial data for the wallpaper makers, the average tax rate for the different professions during the period was just under nine dollars silver money. The most equal group was the painters, the biggest group in numerical terms, with broadly similar numbers for both the average and the median values. For the other professions, the differences were larger with significantly higher average than median numbers.

During the nineteenth century, between 1826 and 1843 (see Table 10.2), the total number of taxed individuals in the painting professions rose slightly, with

Table 10.1 The number of sculptors, gilders, painters and wallpaper makers in
Stockholm who paid ordinary income tax and the average contribution
for each group, 1720–1770 (in dollars silver money)

Year	Sculptors (n)	Sculptors tax	Gilders (n)	Gilders tax	Painters (n)	Painters tax	Wallpaper makers tax
1720	6	19	3	6	9	12	35
1725	10	11.5	3	3.5	10	6	25
1730	9	7	4	6	18	6	9
1735	10	6	2	8.5	17	8	11.5
1740	10	7	4	6	20	8.5	5.5
1745	8	9	3	10	20	8.5	5.5
1750	10	7.5	3	15	15	9	–
1755	12	7.5	6	19	20	9	–
1760	13	6.5	8	5.5	22	9	–
1765	17	6	7	3	28	8.5	–
1770	8	9	5	9	24	10	–

Source: Söderlund (1943), tables 14 and 25.

Note: For wallpaper makers, the number is missing as well as ordinary income tax after 1745.

Table 10.2 The average taxation of sculptors, gilders, painters and wallpaper makers
in Stockholm who paid ordinary income tax, 1826–1843 (in rixdollars)

Year	Sculptors tax	Gilders tax	Painters tax	Wallpaper makers tax
1826	2	4	7.1	5.8
1830	3.3	2	6	6
1835	3.8	2.8	5.3	6.9
1840	5.4	3.1	5.5	6
1843	4.7	3.5	5.5	5.6

Source: SSA, Bemedlingskommissionen, G1c4a, Taxeringslängder.

the exception for the sculptors, a group which lost some of its members. On
average around ten sculptors, ten gilders and ten wallpaper makers were registered as taxpayers each year. The number of painters was more than twice that
number already at the beginning of the period and continued to rise.

The painters on average earned more than the other groups. The incomes
of the sculptors increased whereas gilders experienced a slight decrease in
their incomes. The few wallpaper makers were, in turn, taxed slightly higher
than the painters and substantially higher than the sculptors and gilders.
Taken as a whole, the number of individuals involved in painting professions
was comparatively stable during the period, but slightly increasing from
roughly 45 to 80 in the decades before the mid-nineteenth century. When
compared to the eighteenth century, the average incomes dropped. Painters
by this time were the dominant category both in terms of numbers and with
a significantly higher income level.

Table 10.3 The number of bankruptcies filed by various painting professionals in Stockholm, 1760–1849

Period	Sculptors	Gilders	Painters	Portrait painters	Wallpaper makers	Painter's widows	Sum
1760–1769					2		2
1770–1779	2	1	3		1		1
1780–1789	5	2	4	1	3	1	16
1790–1799	2	3	1	2	1	4	13
1800–1809	4	3	9	1	1	4	22
1810–1819	4	1	7		1	2	15
1820–1829		1	2		5	2	10
1830–1839	2	2	11		5	5	25
1840–1849			10		1	7	18
Sum	19	13	47	4	20	25	128

Source: SSA, Stockholms magistrat och rådhusrätt, C5a, Konkursdiarier.

Table 10.4 The bankruptcy frequency for various painting professionals in Stockholm, 1800–1840, as a percentage of the total number of individuals in each group

Period	Sculptors	Gilders	Painters	Wallpaper makers	Painter's widows
1800	2.7	1.6	1.6	1.3	6.7
1810	5.0	0.6	1.5	2.0	4.0
1820	–	0.9	0.3	3.1	1.3
1830	2.5	2.5	1.6	1.4	2.4
1840	–	–	0.9	0.5	1.7

Source: www.tidigmodernakonkurser.se and SSA, Överståthållarämbetet för uppbördsären-den, Kamrerarexpeditionen, G1BA, Mantalslängder.

For the period 1760–1849, we have identified 128 bankruptcy cases that were filed by various painting professionals. The business sector can be divided into six groups of categories: painters, widows after deceased painters, specialized portrait painters, wallpaper makers, gilders and sculptors. The lowest number of bankruptcies was filed by the gilders, whereas the largest number of cases was filed by the painters, who in numerical terms also formed the largest of the six groups within the business sector (see Table 10.3).

While the painters appear to have filed for more bankruptcies in the two last decades of the period, this was not a true increase. The bankruptcy frequency, here measured as the number of bankruptcies as a percentage of the total number of individuals in each group, in fact, despite certain highs, long term showed a stable pattern at around 1–2 per cent (see Table 10.4).

With few exceptions, the total debts recorded in the 128 bankruptcies filed by the painting professionals were relatively low. Before the 1830s, the average debt was typically below 1,000 rixdollars, after which point, it typically rose slightly above that number. In the 1840s, the sums are relatively

Table 10.5 The average size of the total debt in bankruptcies filed by various painting professionals in Stockholm, 1770–1849 (in rixdollars)

Period	Sculptors	Gilders	Painters	Wallpaper makers	Painter's widows
1770–1779	–	–	862	12,854	–
1780–1789	239	347	163	251	302
1790–1799	270	209	–	–	451
1800–1809	944	173	2,069	1,996	746
1810–1819	525	708	1,023	841	669
1820–1829		192	910	911	285
1830–1839	1,024	3,472	2,143	775	548
1840–1849	–	–	905	1,213	5,132

Source: www.tidigmodernakonkurser.se.

heterogeneous with averages ranging from about 2,000 up to 5,000 rixdollars (see Table 10.5).

Suretyship commitments

The bankruptcy material indicates that the painting professionals had relatively limited credit networks. The number of creditors rarely exceeded ten, even for painters who arguably were the most active group. Even for them, the combined value of the total bankruptcies per decade was comparatively low on average. Their credit networks mainly linked them to low-level artisans and retailers, even if individual wholesale merchants and noblemen also were present. The typical credit relationships were with other painting professionals or guild members. This process crucially also included a strategy of mutual personal sureties.

In the following, the ideas behind the credit networks and the mutual sureties will be detailed and discussed in more detail. We view the disputes that arose as illustrative of the idea behind committed credit commitments, particularly by providing more concrete details about the reasons for agreements than the more general stereotypes formulated by the debtors in their application. We limit the investigation to painters and wallpaper makers, for which a relatively plentiful source material is available.

The painter Daniel Magnus Fernberg, who filed for bankruptcy in 1808, had substantial assets that amounted to 6,632 rixdollars. His debts were higher, however, amounting to 7,143 rixdollars.[2] The largest debt was an unpaid mortgage on two properties that was in his possession, a sum which he owed to Mauritz Aminoff, an army colonel and a member of a Finnish-Swedish noble family. Mortgaged real estate was a common indication of a beginning insolvency in eighteenth and early nineteenth-century Stockholm. The second largest debt of 2,250 rixdollars was to an undefined agency, most likely a state credit agency, a sum which included 270 rixdollars in overdue interest payments. The situation reveals a long-term lending arrangement, which was a common form of money placement in addition to property purchases.

The fourth largest debt of 600 rixdollars was also to a public entity, in this case *Riksens ständers diskontverk*, the main Swedish Discount Bank. Public creditors, like banks and state and municipal institutions and agencies, had claims in most bankruptcy cases. Even though these sometimes were relatively small, including unpaid municipal taxes, they were nearly always actively monitored by the authorities. As opposed to the public credit relationships, the personal credit and associated credit networks built on trust rather than on economic realism and administrative consistency. Often, financial obligations were made between people in the same social circles or with the same professions. This can also be seen in the Fernberg bankruptcy. Among the assets listed in the estate was a claim for 200 rixdollars in the bankrupt estate of a fellow painter, named Westberg.

While the impact of this unpaid debt not necessarily was the main cause for Fernberg's bankruptcy, it certainly played a role. The interesting thing is how such debts directly and indirectly also could affect other members of the same close circles, leading to ripple effects in close-knit social and professional groups.

This can be seen quite clearly when turning to the bankruptcy papers from Olof Liljeberg, another painter, who filed for bankruptcy in 1809, a year after Fernberg.[3] From the letter detailing the reasons for his bankruptcy, Liljeberg in fact lists Fernberg's bankruptcy as the main cause for his insolvency:

> *It would not come as a surprise to my creditors, how I, having trusted my guild brother, the painter Fernberg, as a property owner in this town, to be a wealthy man, which has not made me hesitate in standing surety for him for substantial sums of money, that is now, after his alleged insolvency, demanded from me.*[4]

The specifications show that Fernberg had stood surety for a sum of 975 rixdollars. The sum constituted almost two thirds of the total debt registered in his bankruptcy, suggesting that it indeed formed the basis for his own insolvency. The idea that this, in turn, affected other members of the same professional circle can be substantiated when considering how his assets included money loaned to two other painters, named Holmberg and Mostedt, who owed him a total of 90 rixdollars.

Professional credit relationships were built around personal networks and trust between people. The impact when this trust failed was apparent, also on a personal level:

> *At present, as I am altogether unable to repay the sum in full, and since I have not been the reason for this insolvency, in particular after having for my part not created larger debts for myself than I with certainty have*

considered possible to repay, I humbly hope that my application will be granted and that You, my debtors, after considering what I have here honourably reported, treat me with lenience, in particular, since besides me, a wife and two small children, share the weight of the insolvency that has affected me.[5]

Burghers such as painters and other artisans typically had parts of their assets in the form of claims on other companies in their wider credit networks. Their own credit rating was thus based on the fact that it was possible to recover the claims. Liljeberg's disappointment seemed to have stemmed from the feeling that Fernberg had abused his trust. In reality, a factual assessment of Fernberg's credit rating never seems to have been undertaken or considered.

By stepping in as a guarantor of a loan, the idea of trust was clearly taken one step further than a more typical general financial transaction. As a guarantor, a bankruptcy of a colleague or business partner turned an unpaid claim into more than an uncertain asset. It also meant that the guarantor became responsible for parts of the colleague's unpaid debts. When Liljeberg accepted surety for Fernberg's debt of 975 rixdollars, his own assets became the security for the loan.

Another example of this, seemingly common situation can be found in a bankruptcy from 1835, filed by the painter Johan Fechtonius:

Since I, partly because of sureties, partly for other reasons, have been forced into insolvency, so that I am not at present able to honour my debts, I humble seek to summon my creditors to the honourable Court.[6]

Fechtonius had stepped in as a guarantor for two of his relatives, the farrier-widow Helena Fechtonius and the alderman A. J. Fechtonius. His assets, which far from covered his debts of 720 rixdollars, included various unpaid claims; among other things, an unpaid bill for painting work undertaken for a shoemaker and an unpaid promissory note issued by the painter Dahlström.[7]

The painter Johan Schagerborg, who filed for bankruptcy later that same year, was also forced to do so, due to a guarantee, for a loan of 1,300 rixdollars from *Riksens ständers diskontverk*, given to the brass founder Lindblom. Even though Schagerborg's share of the loan only amounted to 400 rixdollars, his wealth and yearly income was not enough to be able to repay the sum.[8]

Wallpaper makers were also prominent in this kind of credit commitments. In general, relatively few wallpaper makers filed for bankruptcy during the period of investigation. This changed from the late-1810s onwards and in particular during the 1820s, when a chain reaction set in, with multiple bankruptcies in a relatively short time span. Much suggests that the reason for this was the same system of mutual sureties, which also impacted the painters.

Table 10.6 Bankruptcies in Stockholm involving wallpaper makers with one or more surety commitments, 1814–1835

Year	First name	Last name	$\sum$ debt
1814	Carl Eric	Bjurholm	841
1824	Samuel	Lindgren	741
1826	S W	Hagelin	240
1826	Johan Ma	Westman	827
1826	Svante Wilhelm	Hagelin	240
1829	Lars Georg	Öberg	1,834
1830	Johan Magnus	Westman	170
1830	Bror Pehr	Hagtorn	2,461
1835	Johan Eric	Sandberg	305

Source: www.tidigmodernakonkurser.se.

All in all, we have found nine bankruptcy cases involving wallpaper makers, with one or more surety commitments in the period between 1814 and 1835. Such commitments were probably even more common among wallpaper makers than among painters. Much like in the case of painters, the tendency was prominent during the first half of the nineteenth century, but occurred less frequently during the eighteenth century (see Table 10.6).

The detailed information in the bankruptcy applications again makes it possible to show the actual argumentation in many of the cases. This must always be done with some care. When reading the petitions from the debtors, one must keep in mind that the legislation included provisions that provided opportunities for the debtors to find ways out of their predicament. If the insolvency was due to "circumstances", or caused by "ill fate", debtors would in fact be viewed more lenient by the court, be allowed to surrender all property and thus be granted protection from their creditors. If the failure instead was caused by a "reckless lifestyle", by "wastefulness, idleness, fraudulence or negligence", the consequence instead could be imprisonment. It was thus incredibly important for the debtor to present himself in as good of a light as possible.

In the end, the financial information in the files with regard to the economic networks, and the various connections were rarely directly affected by this, however. When filing for bankruptcy, the debtors, under oath, had to provide detailed lists of both their assets and debts and list of the various creditors. When the actual bankruptcy proceedings got underway, the creditors showed up in court and delivered their claims, thus providing further information. When combined, this can be used to paint a detailed picture of the economic situation not only at the time of the bankruptcy application but also of the historical circumstances that had caused the insolvency.

When the wallpaper painter Carl Eric Bjurholm filed for bankruptcy in 1814, his assets amounted to 393 rixdollars, whereas his debts held by 13 different creditors amounted to 841 rixdollars. The largest sum of 400 rixdollars was owed to the Gothenburg Discount Bank. The likelihood that this affected other members of the painting community was very apparent when considering how the loan was guaranteed by the painter Ant. Pousette. When Bjurholm declared bankruptcy, it was Pousette who became liable for the repayment of the debt.[9]

The wallpaper painter Johan Mats Westman filed for bankruptcy in 1826. His assets, as reported in his application, amounted to 616 rixdollars, with debts of 827 rixdollars. The largest debt of 234 rixdollars was to a brewer. Another unpaid loan of almost equally value was in the form of a surety commitment. The loan of 200 rixdollars had been arranged in the beginning of 1826, suggesting its character as a short-term solution to come to terms with a dire financial situation.

This can be further assumed when considering how one of its two guarantors was the fellow painter Svante Wilhelm Hagelin. Westman's complicated economy can be hinted at when considering further information that showed how he, despite his own financial situation, still accepted the role as guarantor for three loans. The total sum of the three obligations amounted to 1800 rixdollar, a substantially higher sum than Westman's own assets, which again suggests that trust rather than economic considerations played a large part in the transaction.[10]

In the context of this investigation, the most interesting aspect of Westman's bankruptcy was undoubtedly the fact that it played a role in the downfall of Hagelin. When the latter filed for bankruptcy later that same year, he implied how this was the result of the breakdown of trust in the agreement:

> *[...] being without funds I received monetary help from my group of friends, who felt my earnest intention to repay them through diligence and work, But as this so far has evaded me, for lack of work, and after I during this period have been unfortunate to enter into guarantor commitments for people, whom I have considered reliable and to be trusted for the debts which I have guaranteed; but found the opposite, when one of them applied for bankruptcy at the honourable Court [...]*[11]

Hagelin revealed assets of 78 rixdollars and liabilities of 240 rixdollars. While his guarantor's commitment in Westman's business seems to have been rather careless, the rest of his finances were not in a better state. Next to his formal debts, he also reported about a further surety commitment for a loan of 200 rixdollars given to a clerk named Weststroöm.[12]

The wallpaper painter Lars Georg Öberg went bankrupt in 1829. His assets amounted to a mere 253 rixdollars while his debts amounted to a total of 1,834 rixdollars. The reasons for his financial state were related in the application, as followed:

> *[...] as a painter in this town, has been unable to avoid incurring several debts, no less in establishing the business than also through subsequent setbacks within my trade, where I have languished or have never experienced*

any success. And notwithstanding the misfortune, which usually happens to beginners in a handicraft, to initially be forced to accept a complete lack of work profits, I have also subsequently, since I became known, not been able to gain any trust [...][13]

Next to the debts and assets in the financial statement, Öberg also revealed no less than three surety bonds to the glazier A. P. Hultander for the sum of 600 rixdollar, to the widow of the tinsmith L. G Lundgren for 300 rixdollars, as well as to the wallpaper painter B. P. Hagström for 80 rixdollars, all three covering loans in *Riksens ständers bank*, the Swedish Central Bank.[14]

The tapestry painter Bror Pehr Hagtorn filed for bankruptcy in 1830. On first sight, his financial situation was not as dire as some of the individuals discussed so far. In the financial statement delivered to the court, he reported a very substantial debt of 2,461 rixdollars. His assets were however also considerable, totalling 2,896 rixdollars. These for the most part were composed of a variety of unpaid bills, including 345 rixdollars owed by a wagonmaker and 253 rixdollars owed by a saddlemaker. His most considerable asset, however, was 750 rixdollars that was connected to a loan in *Riksens ständers diskontverk*. The loan had been originally raised by Hagtorn, but was in the end appropriated by a bricklayer apprentice named Gustaf Ljungström.[15]

The circumstances were explained in more detail in the bankruptcy application. In the text, Hagtorn explained how he had entered the field of wallpaper painting to provide for his family and be an asset to the community. In his view, this call had been hampered by the general financial situation, but also, more notably, by deceitful people:

[...] among whom chiefly must be mentioned the bricklayer apprentice Gustaf Ljungström should be, who have allowed me to become a loanee in Riksens ständers diskontverk for a sum of 5000 rixdollars, through a surety promise signed by him as well as the leather painter Dahlin [...][16]

From the rest of the application, it becomes clear that Hagtorn, despite the surety, never received the money, which in the end forced him seek other solutions. The situation was complicated with claims and counterclaims. Indeed, from the rest of the application, it becomes clear that it was Ljungström, who, according to Hagtorn, together with some of his other creditors, and potentially inspiring further creditors, was trying to restrict his freedom. It was, in Hagtorn's words, the highly spiteful character of this behaviour that ultimately led him to file for bankruptcy to seek protection for his person and family.

The final example to be mentioned here was the 1843 bankruptcy of the wallpaper painter Johan Fredric Berling. The financial statement shows that his assets at the time of the bankruptcy amounted to 413 rixdollars. His debt was dominated by a surety promise, which also was the main cause for his insolvency. The specifics show that Berling together with the gardener Bergstedt had taken on the role as guarantors for a loan of 2,000 rixdollars in *Riksens ständers diskontverk* that had been granted to the provision merchant Fr. Sandberg. Before the third repayment of the loan, Sandberg had declared bankruptcy, which:

> *[…] follows that the rest of the debt in the bank, 1200 [rixdollars] without accrued interest, consequently will fall upon me and me co-guarantor, one for both, and both for one.*[17]

Conclusions

The analysis of the bankruptcies of painters and wallpaper makers hints at one of the limitations of the Stockholm credit market during much of the period of investigation, namely the lack of working capital. While plentiful, private bankers and lenders could not compensate for the lack of government lending or the lack of private banks.

During the period, government credit institutions, mainly various discount banks, promptly demanded repayment of loans when they were due for payment. Among private lenders, renegotiation was often possible as long as interest was paid. Private promissory notes appear to have been transferred indefinitely as long as the holder had creditworthiness. In cases of bankruptcy, discount banks and other government lenders were prioritized creditors and, as such, typically had relatively good opportunities to secure their loans. The result of these recovery activities often triggered a series of bankruptcies when networks of surety bonds were uncovered.

As has been shown, from the 1820s onward painters and wallpaper makers increasingly began using surety bonds to endorse loans, mainly within small clusters of professionals, but also involving other groups in society.

The credit networks were generally limited among painters and artisans in Stockholm, when compared to better-off burgher groups, including merchants, manufacturers or brewers. The individual debts were usually small. The number of debts and claims typically amounted to no more than ten each.

Earlier research has shown that debts to other manufacturers were the most important credit relationship for clothing manufacturers in Stockholm in 1816–1846. While their businesses were usually on a much larger scale than that of painters and wallpaper makers, an in-depth study of two manufacturers and a textile retailer in the 1830s shows that the trio supported each other heavily through mutual loan guarantees. When one filed for bankruptcy, the others were therefore also included in the bankruptcy.

We have now seen that this was a reoccurring behaviour among painters and wallpaper makers as well. There are changes however. Neither the share of the debts nor the systematic nature of the surety bonds reached quite the same levels or had the same impact among painters and wallpaper makers, as has been previously shown for clothing manufactures.

The results from the investigation indicate that the early modern notion of creditworthiness was slowly beginning to be replaced by higher demands for collateral in the credit market in Stockholm after the reform of the Swedish bankruptcy institution. We view the growing use of promissory notes with collateral as a kind of modernization compared to creditworthiness based only on general conceptions of the borrower's ability to pay based on, for example, titles or social position. The debtors' arguments can in contrast hardly be described as "modern" or "rational" in a strictly economic sense. In several cases, painters and wallpaper makers made guarantees that far exceeded their assets. This applies not least to public loans, which were both extensive and typically also given priority when banks pursued their claims in the event of a bankruptcy.

A summary of the results suggests that the reform of the bankruptcy institution in Sweden between 1767 and 1818 tightened the requirements for security in the credit market. Even so, these new financial instruments were handled in the context of the remaining early modern notions of creditworthiness and confidence, well into the nineteenth century.

Notes

1 For a more detailed treatment of the topic, which also forms the basis for this chapter, see Hagberg and Nyberg (2017a, 2017b).
2 SSA, Stockholms magistrat och rådhusrätt, F6a, Konkursakter, 60/1809.
3 SSA, Stockholms magistrat och rådhusrätt, F6a, Konkursakter, 116/1810.
4 "Det lärer icke vara mine Herrar Borgenärerna obekant, hurledes jag, i förlitande på min Embetsbroder, Målaren Fernberg, så som fastighetsägare här i staden, vore en förmögen man, igrund varaf jag icke kunnat draga i betänkande, att för honom ingå Borgen för betydliga penningsummor, hvilka numera, efter hans förmodligen inråkade obestånd, af mig utkräfvas".
5 "För det närvarande aldeles ur stånd satt, at här före kunna åstadkomma full betalning, samt enär jag sjelft icke varit vållande till detta mitt obestånd, hålt jag för egen del icke åsamkat mig större skuld, än den jag med säkerhet ansett mig kunna betala, vågar jag ödmjukast hoppas att denna min ansökning bifalles samt att mina herrar Borgenärer vid övervägandet af hvad jag således haft äran undraga icke torde med mig på det strängaste förfara, hållt jämte mig, en hustru och tvenne små barn dela tyngden af det obestånd som mig drabbat.".
6 "Sedan jag, dels genom ingångna borgens förbindelser, dels genom andra orsaker, blivit så bragt på obestånd att jag för närvarande icke är i stånd att honorera mina skulder, få jag hos högädle Rådhus Rätten ödmjukast att mina herrar Kreditorer få sammankalla".
7 SSA, Stockholms magistrat och rådhusrätt, F6a, Konkursakter, 119/1837.
8 SSA, Stockholms magistrat och rådhusrätt, F6a, Konkursakter, 121/1838.
9 SSA, Stockholms magistrat och rådhusrätt, F6a, Konkursakter, 110/1816.
10 SSA, Stockholms magistrat och rådhusrätt, F6a, Konkursakter, 109/1828.
11 "[...]såsom medellös erhöll jag penningaförstreckningar till mitt etablisement af vänner, vilka kände min oryggliga föresats, att genom flit ioch arbete dem gottgöra, Men som detta har hitintills mig ej lyckats, i brist undstående arbeten, samt att jag under denna tid tillika varit nog olyckligt att ingå prioerie Borgensförbindelser för andre, hvilka jag ansett säkre och vederhäftige för de skulder jag för dem borgat och funnit motsatsen, då en af dem hos Högädle RådhusRätten sökt Cession [...]".

12 SSA, Stockholms magistrat och rådhusrätt, F6a, Konkursakter, 74/1827.
13 "[...] såsom målare häri staden, ej kunnat undvika att åsamka mig åtskillige skulder, ej mindre för mitt etablisement uti rörelsen, än ock genom sedermera inräknade motgångar inom mitt yrke, der jag förga eller intet fått erfara någon lycka. Och oberäknad det misöde, som vanligen händer nybörjare af handtverk, att i början nödgas sakna all arbetsförtjenst, har äfven sedn jag blev känd ich kunde tillvinna mig något förtroende [...]".
14 SSA, Stockholms magistrat och rådhusrätt, F6a, Konkursakter, 74/1830.
15 SSA, Stockholms magistrat och rådhusrätt, F6a, Konkursakter, 591/1831.
16 "[...] hvaribland hufvusakligen må nämnas Murargersällen Gustaf Ljungström, hvilken inlåtit mig såsom låntagare uti Riksens Ständers BancoDiskon för en summa 5000 rd banco, med borgen tecknad af nämnde Ljungström tillika med Lädermålaren Dahlin [...]".
17 "[...] följer att den återstående Bankodiskontskulden 1200, utan uplagne löpande räntor, således komma att drabba mig och medlöftesman, en för bägge och en bägge för en [...]".

Part IV

Conclusions

11 The Stockholm credit market in an international perspective

Klas Nyberg

Introduction

What can the transformation of the credit market in Stockholm and its impact on luxury manufactures and artisanal crafts during the transition to modern times teach us when viewed in a European comparative perspective? In the following the results of the anthology will be discussed and contextualized under four headings in relation to previous research about the problems that has been investigated.

An attempt will be made to make an overall assessment which gives a partly new picture of Stockholm's credit market and the people that worked within its boundaries, against a wider European background and in the light of the demographic and economic stagnation of the city between 1760 and 1850.

The case studies have shown that the transformation of the credit market and the credit conditions in the various fashion and luxury industries mainly were affected by:

I The impact of the stagnation of Stockholm between 1760 and 1850, where companies in the textile and artisanal industries in Stockholm showed an inability to expand as market involvement increased after 1760 despite the gradual liberalization of the Swedish economy from the late eighteenth and in the beginning of the nineteenth century.

II The institutional design of manufactures, the guild system and credit institutions after 1739, through which an interaction was created between hallmark courts and the guild system with special credit institutions for the financing of textile and metal manufactures as well as for parts of the furniture manufactures (Chapters 5, 6, 9 and 10). The same kind of institutional structure that seemed to inhibit Stockholm functioned dynamically in other parts of Sweden (Gothenburg and Norrköping).

III The local credit market in Stockholm, which showed industry-specific patterns in the fashion and luxury industries. The interpretation suggests that the success of entrepreneurs and business owners in the silk weaving industry, among wig makers, furniture makers and painting professionals was due to the extent that the different types of consumer goods and services that they provided were adapted to Stockholm's stagnant market, or also worked in the growing consumer goods markets in the countryside (Chapters 6, 7, 8, 9 and 10). We can see a more complex use of credit in Stockholm with an increased demand for collateral among producers and traders in Stockholm from the 1820s, whereas the development in the rest of the country saw an increased use of promissory notes against name security only, during the first half of the nineteenth century.

IV The reformation of the bankruptcy institution that was undertaken between 1767 and 1818 (Chapters 3 and 4). The development led to an increased efficiency with a gradually decreasing number of days from application to verdict. In Stockholm the development can be linked to a continuous increase in the number of cases after the reformation period. In Gothenburg, the number of cases on the other hand decreased during the same period.

Did these four problem areas constitute a specific Swedish development pattern linked to the different conditions in Stockholm and growing factory and trade towns in the rest of country, or did it form part of a more general European development?

"A stagnating metropolis" revisited

The development of the Stockholm credit market and luxury industries must first and foremost be understood in light of the fact that its economy was one of Europe's weakest urban growth areas, together with trading cities in southern Europe and the capitals Amsterdam, Copenhagen and Lisbon.[1]

Stockholm was among the long-term stagnant European cities that emerged in the eighteenth century. The city was on the northern outskirts of a European periphery, where Italian, Spanish and Portuguese cities in the Mediterranean made up the southern. In this sense, the northern and southern borders shared common characteristic with a stagnant demographic and economic development from the second half of the eighteenth until the first half of the nineteenth century.

This both absolute and relative decline has been explained by the decline of the Baltic and Mediterranean trade and the corresponding rise of the Atlantic trade. The shift in the economic centres in Europe was also connected to the industrial development. The expansion of British and subsequently Belgian industrial cities created and expanded a north-west European centre of industrial and economic growth. It was the same geographical area that Jan de Vries put at the centre of his consumption-dynamic theory of an industrious revolution from the end of the seventeenth century.

The development of the urban credit markets is not as clear-cut. Here, there was a clear northern shift of gravity from Amsterdam to London, and to Hamburg, which also was important for Stockholm. Most of these important commercial centres were capitals which were generally "stagnant" to a greater extent than the British and Belgian industrial cities. According to Söderberg, Amsterdam, Copenhagen and Lisbon show the same kind of stagnation as Stockholm.

In Sweden, Stockholm lost its industrial lead to the factory city of Norrköping and with regard to foreign trade to the trade and industrial city of Gothenburg (Chapter 4). In many respects, the capital still remained a financial centre, however.[2] Population growth related to the level of urbanization and the expansion of textile manufacturing and crafts. The dynamics of early urbanization, the "urban-industrial growth poles", were mainly in western Sweden, where market-orientated, farmer-driven agriculture expanded greatly, while the large-scale manor-based agriculture in the Stockholm region stagnated.[3] A Swedish industrious revolution probably had its starting point in Western Sweden, but it was a process that was significantly delayed compared to the development in Northwestern Europe, in the Netherlands and Great Britain.

The reason why the capital remained a financial centre was likely Stockholm's historical role as a "contact zone" in the words of Kapil Raj. The city had been the centre of power and trade since the seventeenth century with special legislation for both trade and manufacture, which unilaterally favoured it at the expense of the country's other staple cities. This included key trade legislation that required all ships sailing to or from ports in the northern part of the country to pass through Stockholm. It was Axel Oxenstierna, the influential Chancellor of the Realm, who coined the famous original slogan about the capital's special character already in the 1600s: "*Man skall all concilia därhän dirigera* [...]", with the implication that if the economic policy is designed so that trade and industries are concentrated in Stockholm, they will flourish in other cities as well. The same ideas could be seen in several other European countries in the early modern period. During the period, the leading elites, including the nobility and the leading burghers, built palaces in the capital, including at *Skeppsbron*, the easternmost part of the centre of Stockholm. Directly overlooking the harbour, the area became a familiar silhouette which met ships arriving from the Baltic, carrying new goods from western markets (Chapter 5).

The investigations of the credit networks for wig makers and other artisans that filed for bankruptcies show that they were mainly involved in local credit networks. Therefore, the long-term economic downturn is likely to have had a strong negative impact on their economy when compared to silk manufacturers or furniture makers who also sold their products on the national market. When rates of bankruptcies rose, the use of sureties reflect that the requirements for security also grew in certain groups from the 1820s. The local economic sphere usually lacked credit relationships with both national and international players.

It is worth mentioning that a type of protected business areas emerged in Stockholm during the later phase of the stagnation, in the first decades of the nineteenth century. In these areas, female professionals were shielded from male competition, a development with parallels in several European countries, especially in Paris and Northwestern Europe.[4] The development in Sweden happened at around the same time as a growing male mortality rate, which Johan Söderberg already in 1991 placed in connection to the subsequent rise of co-habitation without marriage, the so-called "Stockholm marriage".[5] This development, in turn, was linked to the rise of a growing number of female businesses, including in fashion-related retailing after the end of the Napoleonic Wars. In recent research, this has been shown to have been a more general Northern European phenomenon. It was a new kind of way of living that happened in parallel to increasing female business activities. In Stockholm, business sectors exclusively reserved for women were created, including small wares sales and overwater transportation. From the late eighteenth century, a more market-oriented transition to a division of women's and men's tailoring also emerged, where seamstresses much like in Paris took over the former branch, which previously had been dominated by male tailors.[6]

There is much to suggest that the international trade crisis in 1763 and the meeting of the Swedish Diet in 1765–1766 with a shift in the political majority must be seen as a watershed moment in the development of Stockholm's role as a production site for luxury textiles and artisanal handicraft. The policy of favouring the capital with a special economic legislation, which intensified in the 1720s and 1730s, after the end of the Great Northern War, culminated in an economic and social crisis in the 1760s, triggered by the 1763 trade crisis. This led to social and political debates with demands for, among other things increased financial freedoms and improved state finances, where the extensive subsidies given to the textile manufacturers were considered harmful. This came to a head at the meeting of the Diet in 1765–1766. In its wake, a deregulation of the Swedish economy was introduced, in a series of reforms that continued well into the nineteenth century. The manufacture policy was changed, with the removal of subsidies and the introduction of a much more restrictive state loan regime. The previously regulated grain trade was opened up completely in 1775; a reform of the rural market system was introduced in 1788; the town tariff on goods imported from the countryside was abolished in 1810 and the trade in foodstuff was deregulated in 1815. The list of reforms, all of which were designed to facilitate the exchange of goods in the Swedish economy, can be made much longer. The unilateral support of Stockholm at the expense of other cities had been one of the major obstacles to a free market exchange.[7]

The reform pace was not fast and in the short term could not counteract the stagnation, but instead initially seems to have reinforced the negative development. The period was marked by rising mortality rates, sharp price rises and social and political unrest that culminated with the assassination of Gustav III in 1792 and the murder of Axel von Fersen, the Marshal of the Realm by a mob in 1810. During the period trade, manufactures and other industries in the city experienced considerable difficulties and faced a long-term decline.

The fundaments in the economic legislation thus undoubtedly played a role in the stagnation that affected the capital after the middle of the eighteenth century. There is also reason to suspect that the part of the economic policy that favoured Stockholm had a negative impact on the development in the rest of Sweden. This include regulations such as the above-mentioned legislation that channelled the trade to and from northern ports through Stockholm as well as mercantilist measures modelled on the English Navigation Laws that stipulated that foreign vessels only were allowed to import goods from their own countries or its colonies to Sweden. This also applied to the state support system for manufactures as well as to the prohibitions on textile imports. All of these regulations were intended to benefit Stockholm directly or indirectly, but probably added costs to both trade and production. Were in fact business owners and entrepreneurs in Stockholm disadvantaged by their favoured status and was perhaps the stagnation an effect of this?

Early modern institutions in the fashion and luxury industry

The transformation of the credit market in Stockholm was greatly affected by institutional factors where the creation of a state-supported manufacture system became the most important in context of fashion and luxury. Sweden was one of the countries in Europe where manufactures were encouraged through a specially designed national support structure with targeted credits. This provided the fundament for the creation of a system where local manufactures helped in the spread of fashion in the field of textile and artisanal handicraft through a process of domestic imitation and local production, made possible through import restrictions and tariffs. This contrasted with other countries on the periphery where fashion impulses spread solely through the import of fashion goods from the industrial leaders in the centre or through sumptuary laws (Chapter 5).

The development of the institutional structure in eighteenth-century Stockholm can be interpreted with the help of Sheilagh Ogilvie's institutional approach on the role of institutions in parts of Europe during early modern times. Ogilvie's overall theoretical approach of state corporatism with its focus on the interplay between different kinds of guilds in trade and production has renewed the debate of early modern institutions in a fundamental way. In several of her works, Ogilvie analyses the question of the importance of the early modern proto-industry and the guild-system for regional economic development. She not only discusses urban craft guilds but also analyses other forms of the guild system, including merchant guilds, both on their own and in relation to the municipal authorities and the early modern absolute and/or feudal state.[8]

Her case studies focus in part on the production of worsted cloth. The institutional and social arrangements in Europe also varied with regard to the textile processing method of "new draperies" from the end of the sixteenth century onwards as well as traditional woollen cloth. A general trend which is stressed in earlier research into the woollen and worsted textile industries is that the guilds lost influence over time as merchants strengthened their grip through their influence in town councils. This trend is most prominent in the production of woollen cloth. In the towns of late medieval Italy and Flanders, fulling, shearing and dyeing were processes that were undertaken by experts who were organized in traditional craft guilds. By the end of the early modern period, these guilds had often been overshadowed by merchants who increasingly controlled the preparation and sales of the textiles. In Stockholm, a similar development happened in the eighteenth century, when the hallmark court was given a similar function when clothiers, dyers, shearers, furniture makers and wallpaper makers were placed under the administrative control of the new manufacture legislation.[9]

Ogilvie's analysis of institutions as barriers is closely related to the institutional structures that developed in Stockholm after the end of the Great Northern War. The new institutions that governed textile manufacturing and artisanal crafts added to the ongoing stagnation process. In Stockholm, they locked the textile manufactures in procedural disputes. In the wool city of Norrköping where textile manufactures were controlled and owned by merchants and in the metal manufacturing city of Eskilstuna the same kind of control in contrast seems to have been supportive and dynamic. One of the results of this was that the rural factory towns expanded in the early 1800s whereas Stockholm was de-industrialized (Chapter 5). The stagnation of the manor-based agricultural production in the Stockholm region as opposed to the success of the dynamic western Swedish farming economy is a direct parallel to this development.[10]

Several studies of the development and decline of industrial regions in various countries during the nineteenth century have shown that industrial environments similar to Norrköping with market-savvy producers were more successful in the long term than areas where the manufactures transferred sales to independent merchants. R. G. Wilson in his studies on the role of merchants in the British woollen industry cluster in West Riding in Yorkshire in the late eighteenth century states that the main reason for the concentration was a group of specialized traders with a background as manufacturers. Their main strategy, as in Norrköping, was to focus production on medium-quality qualities for exports. By managing to control also the subsequent stages in the finishing of the textiles after they had left the mills by employing small businesses in cities such as Leeds and Halifax, a small-scale rural production structure was maintained. The traditional woollen regions in the western parts of the country and in East Anglia, to a greater extent rested on an indirect structure that involved merchants in London, were less successful in adapting to market demands and declined in the long term.[11]

Pat Hudson likewise has shown how institutional arrangements could support different forms of production during the early industrialization. In the West Riding wool industry, during the transition to industrialization, for example, small-scale producers supported clothiers with joint solutions to handle the later processes in the textile production, through so-called "company mills".[12]

When compared the situation in Sweden, it appears likely that the Stockholm's textile luxury industry (Chapters 2, 5 and 6) lacked this ability to adapt and instead declined when competition intensified in the early 1800s. Instead, the trading town of Gothenburg and the factory town of Norrköping expanded.[13]

One of the central aspects to affect the development of the fashion and luxury goods manufactures was the development of institutions that tried to solve problems related to the financing of the industry. Fiscal experiments in the Manufacture Office, the new overseeing body established in 1739 to handle the administration of the manufacture system, played a crucial role. The emergence of many controversial forms of support had been preceded by long-standing discussions during parliamentary debates in the previous decades.[14] In this process, the idea of short-term lending against sureties from the Manufacture Discount Bank, a special business support bank set up to support the domestic industry was introduced. Between 1739 and 1766, the Manufacture Office distributed state support through the Manufacturing Fund, which was the forerunner to the later Manufacturing Discount Fund.[15]

Over time, the lending process was modernized. From the 1750s, loans against mortgage were gradually replaced by loans on promissory notes. Most often, this was debt securities obtained by trading retailers when selling goods on credit. From this activity, the Manufacture Discount Bank was developed as a standalone fund. After the Manufacture Office was closed down after the Diet of 1765/1766 the business in a heavily revised form was transferred to the Board of Commerce, the old state central administrative department. The subsidy was now instead focused on low-interest rates that undercut the 6 per cent rates on the usual credit market, but the subsidies were not enough for the textile manufacturers need for discounts. The reason for the limited lending ability of the Manufacture Discount Bank and hence its inability to meet the demands of the companies was the limited credits which it was given by the Swedish central bank. Nevertheless, the Manufacture Discount Fund became one of the most important and lasting public lenders and, just like private discounts, played an important role in the credit market.[16]

The Manufacture Discount Fund had been set up mainly to solve the basic financing problems in the textile industry caused by the relationship between commodity credits and sales credits. Virtually, all purchases and sales of goods and services in the textile industry before 1850 took place on credit with varying maturities and interest rates. For the producer, the basic financing problem was that sales credits had a duration (9–11 months) that was three times as long as the duration of the credits that he himself had to agree to when purchasing raw materials (about three months). To this should be added the costs of salaries. Textile manufacturers as a result suffered a constant shortage of capital and were threatened by liquidity crises. Pat Hudson's aforementioned study of the financing problems of the British wool industry shows that one must know the structure of the capital needs in order to be able to interpret public and private credit in the overall financing of individual industries. The industry's capital requirements changed over time as did different types of business.[17]

The long sales credits are also the problem most often mentioned in the contemporary information from protocols, submissions and reports about the state of the industry. One such report stated that the capital required in the business was equivalent to the double value of the finished goods to be able to pay wages and sales credits, when the fixed capital was unaccounted for.[18] Problems in the sales process were also allegedly inhibiting further expansion of the business. In addition, in order to secure a transaction with a 12–18-month credit at all, the manufacturer had to sell without an actual promissory note, which normally functioned as collateral.[19] Internationally, the need for capital was influenced by whether the textiles were exported, as in West Riding, or if the sales were limited to the domestic market, as in Scandinavia.

The character of credit in the various business sectors

Part III of the anthology dealt with a number of business and industrial sectors during the eighteenth and early nineteenth century. One group included studies of silk manufactures (Chapter 6), furniture makers (Chapter 9) and painting professionals (Chapter 10), three areas which after 1739 for the most part, but with some exceptions were answering to the hallmark court system. Another study focused on hair professionals (Chapter 7), who developed as a profession in Sweden from the seventeenth century as part of the French impact on Swedish fashion. Hair professionals engaged in a special type of artisanal handicraft in which they combined craftsmanship with trade, partly buying and trading hair and partly engaging in the fashioning of the hair into wigs, as part of an elaborate manicure. A final study examined the role played by book printers (Chapter 8). While in some sense differing from the other industries book printers indirectly constituted a central component in the luxury and fashion context that has been investigated here. They were both the most important publishers of the first fashion magazines and thus acted as mediators of notions of luxury and fashion, and were also, at least partly, forerunners in the emergence of advertising which gradually became the most important component in the spread of fashion to a wider middle class across the country. As shown in the study of the French influence on Swedish fashion through fashion magazines and advertising, it was in the daily press that the fashion designations that first appeared in fashion magazines were linked to textiles and other accessories that retailers marketed as fashion goods (Chapter 2).

Håkan Jakobsson has investigated the silk weaving industry and its endurance and flexibility during a period that covers the 1760s, as seen one of the defining moments in the development of both manufactures and the start of a stagnating economic development in Stockholm. The study can be seen as a follow-up to the investigation by Per Nyström who in 1955 showed how the silk industry managed to adapt after the 1760s through measures, including a shift in its production. This case study was a central post in what has come to be called the Heckscher–Nyström debate where Nyström challenged Eli Heckscher's interpretation of the manufacture system as a whole and in particular his view of the textile manufactures in Sweden as "artificial greenhouse flowers" that depended on government subsidies for survival.[20] Jakobsson's study has shown how the silk industry in Stockholm functioned as a cooperative organization where businessmen, traders and bankers intermingled. This network-based grouping of entrepreneurs, partly organized in a trade society with community facilities similar to the "company mills" in Yorkshire, seems to have contributed to different kinds of flexibilities. The transfer of property, clustering and activities to reduce the threat of bankruptcies provided continuity in the largest companies. The strategies also enabled the businesses to quickly reduce their scale during crises and then expand again in favourable economic conditions, a direct parallel to the "flexible specialization" theory previously applied to studies of Lyon's silk industry.[21] Jakobsson also points out how the Swedish silk industry built on a strong French model and in its early phase made use of French expertise. It was thus also one of the imitators that merchants and manufacturers in the European silk centre Lyon sought to counteract through export restrictions, and other barriers, and by adapting special fashions that would be suitable for the German, British and Spanish markets.[22] The Swedish silk production did not follow the international fashion trends, which, among other things, Natalie Rothstein has identified for the period after 1770. When the international output remained more diversified, the Swedish production increasingly focused on a smaller range of simpler qualities[23] (Chapter 6).

Göran Ulväng has investigated the credit conditions for furniture makers during bankruptcies and Axel Hagberg and Klas Nyberg the similar situation for painting professionals. The increasing consumption of new luxury, including art, decorative elements and artisanal handicraft, during the period is reflected in the growing number of individuals who worked in specific painting professions in the early 1800s. At the same time, the number of practitioners in other related fields, including sculptors, gilders and wallpaper makers, decreased. This difference should not be linked to differences that have to do with the stagnation of Stockholm or the expansion of rural markets. Both of these groups were in fact active on the same urban market. The number of carpenters instead decreased after 1790 but from a high level, with about 90 active carpenters some decades before. The number of chair makers also decreased from 1760 to 1820 but thereafter saw an increase (Chapters 9 and 10).

The study of painters and other bankrupt artisans suggests that the lack of working capital, one of the weaknesses of the credit market, in particular in the case of painter and wallpaper makers, limited the further growth of their numbers. They did for example not have access to the various credit institutions that supported the textile manufactures. We have seen how, from the 1820s onwards, people in the painting professions were actively using guarantor connections that lacked coverage. Painters and more so wallpaper makers stood surety for each other's loans within small clusters of colleagues and friends, but also involving other groups. Similar patterns in the wool industry indicate that this was an important development in the de-industrialization of Stockholm during the stagnation period.[24] In this sense, the furniture craft was more sustainable by leaning on orders from owners of manor and ironworks in rural areas in the Stockholm hinterland (Chapter 10).

Mats Hayen's contribution draws attention to the relationship between the emergence of mechanized letterpress printing in Stockholm and fashion. In the bankruptcy wave that followed the early establishment in the 1830s, most individuals within a fairly limited cluster appear to have been involved as creditors and debtors within an important but rather limited financial network.[25] Hayen's analysis should be seen against the background of studies on the development of the early media, which testify to a similar development as early as the late eighteenth and early nineteenth centuries. The emergence of periodic publications (daily press, newspapers and magazines) was an important part of this development. In Great Britain, during this period, special fashion magazines for men and women became more numerous. The information was disseminated into the countryside, mainly with the help of travelling traders. Rural newspapers emerged in the eighteenth century. The newspapers that were established before the mid-nineteenth century were founded almost exclusively by actors active in the book market. In cases where bookshop owners and/or book printers did not establish the newspapers, such actors often took on a role as publishers later in the newspaper's history. The emergence of the printing press was thus absolutely central to the mediation of fashion and the emergence of advertising of consumer goods at the beginning of the nineteenth century (Chapter 8).

The Swedish domestic market for luxury goods expanded beyond Stockholm during the period of investigation, even though the development saw wide regional variations. Support for de Vries' theory about an increasing industrious behaviour had been provided for Swedish circumstances already in 1979 when Lennart Schön argued that the expansion of textile manufacturing before the mid-nineteenth century was part of consumer dynamic development.[26]

The need for consumer credit most likely increased in the light of this increased demand for consumer goods, luxury and fashion. The concentration of printed information, wig makers, textile manufactures and artisanal crafts of an increasingly international standard made Stockholm the centre of luxury consumption in Sweden and the place from where the notion of new luxury was disseminated to the countryside, where an industrious revolution led to increased and broader demand. The capital was the place where fashion magazines were produced and the most developed daily press existed as well as where specialized retailers for fashion goods and the majority of the wealthiest consumers resided. Was perhaps this concentration not altogether positive, and in the end only favoured competitors who were active in less competitive cities such as Norrköping, Eskilstuna, Malmö and Gothenburg?

The study of painters and wallpaper makers (Chapter 10) and two previous surveys of clothing manufacturers support a picture of growing difficulties to survive in growing competition. As mentioned, the rising bankruptcy rate in Stockholm from the 1820s is closely associated with growing demands for loan securities. These claims appear to have been raised by institutional lenders, but it is uncertain to what extent this development increased security requirements in general at this time. It is feasible however that the loan guarantee requirements not only reflected conditions on Stockholm's local credit market or was a reflection of the prolonged economic stagnation in the capital but also a growing lack of ability to compete with regional markets.

Worst, perhaps, were the conditions for hair professionals, who in comparison with many other artisans stood out as an exclusive craft connected to and highly dependent on the consumption of an urban elite already from the end of the seventeenth century when the French wig fashion was fully realized in Sweden. Initially, they were active participants in the credit market in Stockholm, both as financiers and debtors. In the long run, however, they were affected by the change of hair fashion and slowly faded away into increasing poverty when the demand for wigs waned (Chapter 7).

The reform of the bankruptcy institution, 1767–1818

The increased need for credit with the emergence of short-term unsecured loans in parallel with a growing number of bankruptcies prompted a need for a reform of the bankruptcy institution in Sweden. The process developed after the crises of the 1760s and continued in the first decades of the nineteenth century.

The overhaul of the existing legislation and introduction of new laws took place in an international context of reform approximately at the same time as in several other European countries. In Kapil Raj's terminology the development matches the bureaucratic practice that was introduced by state and municipal officials. During the eighteenth century, the legal influence in Sweden was mainly from the German-speaking world, but was subsequently replaced by French influences in the nineteenth century (Chapter 3).

Stockholm in this respect also functioned as a contact zone, again in accordance with Kapil Raj's terminology.[27] In the eighteenth century, French aesthetic ideals, a German-speaking bureaucracy and foreign merchants dominated. Immigrant elites acted as intermediaries in several important respects, including through their involvement in international trade, the transmission of French aesthetic ideals in architecture, arts and fashion, as well as in the industrial legislation and the various institutions that affected the manufacture system, which also drew inspiration from French models (Chapter 2).

The reformed bankruptcy institution helped to modernize the broader credit market in two steps. First, the reform of the legislative process meant that bankruptcy cases were settled faster, on average twice as fast. Second, there was a development especially in the Stockholm credit market, where security requirements increased among artisans and textile manufacturers when compared to the eighteenth century (Chapters 9 and 10).

The use of surety commitments also increased during the 1820s and 1830s among the artisans surveyed. A similar tendency has previously been shown to have occurred in Stockholm's textile manufactures between 1816 and 1846. Frequently, the guarantor here lacked financial cover for his surety commitment and went bankrupt himself if the borrower filed for bankruptcy. This new economic behaviour should be seen as one of the partial explanations for the increased number of bankruptcy cases in Stockholm after 1820. It can be interpreted as the bankruptcy institution modernized the credit market, but that manufacturers and artisans partly retained an older behaviour with early modern roots characterized by a lack of economic realism during this transition period. Stockholm's development with an increasing number of bankruptcies from the 1820s onwards was not a general tendency, however. In the trade city of Gothenburg, the number of bankruptcies decreased slightly, and the same also applied to the industrial city of Norrköping (Chapter 10).

Notes

1 Söderberg et al. (1991), ch. 2.
2 Nyberg (1999, 2009).
3 Nilsson (1989); Söderberg et al. (1991), p. 208.
4 Crowston (2001); Coffin (1996); Simonton and Montenach (2013); Simonton et al. (2015); Ilmakunnas et al. (2018).
5 Söderberg et al. (1991). For the concept of "Stockholm marriage" see Matović (1984).
6 Bladh (1991, 2008); Rasmussen (2010).
7 Lindström (1923, 1929); Staf (1940).
8 Ogilvie (1997, 2003, 2019).
9 Nyström (1955), ch. 1.
10 Söderberg et al. (1991), p. 206.
11 Wilson (1973); Smail (1999).
12 Hudson (1986).
13 Nyström (1955); Schön (1979); Persson (1993); Nyberg (1999b).
14 Already between 1720 and 1739 more than one million dollars silver money was distributed as support from a National Aid Fund. Fritz (1967).

15 Fritz (1967); Gårdlund (1947); Nyberg (1999b).
16 Fritz (1967), ch. III.
17 Hudson (1986).
18 Norrköpings stad, landshövdingeberättelser, 1832–1836, p. 53.
19 Norrköpings stad, landshövdingeberättelser, 1837–1841, p. 33; Stockholms stad, landshövdingeberättelser, 1838–1842, p. 20.
20 Heckscher (1932, 1937/1938); Nyström (1955); Nyberg (1992).
21 See Nyberg (1992) for similar observations about the wool manufactures in the 1780s, and the idea of "flexible specialization".
22 Poni (1997), pp. 57, 71.
23 See Rothstein (1990); Nyström (1955); Nyberg and Ciszuk (2015).
24 Persson (1993).
25 See Rebolledo-Dhuin (2016) for bankruptcies in the Parisian book trade during the same period.
26 Schön (1979), ch. III.
27 Raj (2011).

Bibliography

Archival sources

Riksarkivet (RA)

Kollegiers m. fl:s skrivelser till Kungl. Maj:t

Kommerskollegium
  Huvudarkivet
    Förteckningar över utfärdade fabriksprivilegier

  Advokatfiskalkontoret
    Årsberättelser

  Kammarkontoret
    Årsberättelser fabriker serie 1

Ericsbergsarkivet
  Autografsamlingen

Eli Heckschers arkiv

Stockholms stadsarkiv (SSA)

Stockholms magistrat och rådhusrätt
  Konkursdiarier
  Uppropsprotokoll
  Lagfarts- uppbudsprotokoll
  Konkursakter, huvudserie

Rådhusrättens 1:a avdelning (Justitiekollegium, Förmyndarkammaren)
  Bouppteckningar

Byggnadsnämndens expedition och stadsarkitektkontor
  Bygglovsritningar

Handelskollegiet
  Borgarböcker, huvudserie

Katarina kyrkoarkiv
   Död- och begravningsböcker

Storkyrkoförsamlingens kyrkoarkiv
   Lysings- och vigselböcker, huvudserien

Stockholms stadsingenjörskontor
   Förrättningsprotokoll
   Designationer och planritningar

Överståthållarämbetet för uppbördsärenden
   Kamrerarexpeditionen
   Mantalslängder

Landsarkivet i Göteborg

Ämbetsarkivet
   Övriga register

Dictionaries and Encyclopedias

Dictionnaire universel françois et latin vulgairement appelé, Dictionnaire de Trévoux.
   (1771). 8 vols. Paris: Vincent.
Le Dictionnaire de l'Académie Française (1694), 2 vols. Paris: Jean Baptiste Coignard.
*Encyclopédie ou Dictionnaire raisonné des sciences, des arts et des métiers, par une
   société de gens de lettres* (1751–1777), (eds.) M. Diderot and M. D'Alembert. 35
   vols. Paris: Briasson.

Periodicals

Cabinet des modes ou les modes nouvelles (1785–1793). Paris.
Gallerie des modes et costumes français (1778–1787). Paris.
Journal för Konster, Moder och Seder (1815–1818). Stockholm.
Journal de Paris (1777–1840). Paris.
Konst och Nyhets Magasin för Medborgare af alla Klasser (1818–1823). Stockholm.
L'Avantcoureur (1760–1773). Paris.
Le magasin des modes nouvelles, françaises et anglaises (1786–1789). Paris.
Magasin för Konst, Nyheter och Moder. En månadsskrift (1823–1844). Stockholm.
*Tableau général du goût, des modes et costumes de Paris par une société d'artistes et
   gens de lettres* (1797–1799). Paris.

Printed sources

Arnell, L. (1825) *Huru skola Bankrutter förekommas, Wingleriet och Ockret utrotas?.*
   Stockholm: Carl Deleen.
Berch, A. (1747) *Inledning til almänna hushållningen, innefattande grunden till politie,
   oeconomie och cameral wetenskaperne.* Stockholm: Lars Salvius.
Berg, L. O. (1969) "Hall- och manufakturrätten i Stockholm. Anteckningar om dess
   bakgrund, historia och arkiv", in *Stockholms stads arkivnämnd och stadsarkiv*

årsberättelse, pp. 45–72. Also reprinted in Stavenow-Hidemark & Nyberg (2015), pp. 49–60.

Bergström, C. (1771) *Tankar om Svenska Hushålls-Lagfarenheten*. *Om gäldbundna personers otillräckeliga egendoms afhändande, till flera borgenärers förwaltning och betalning*. Uppsala: Johan Edman.

Bihang till Riks-Ståndens Protokoll vid urtima Riksdagen i Stockholm 1817 och 1818. Nr 18. Stockholm: Zach Hæggström

Björkman, R. (2004) *Skräddarlexikon*. Stockholm: Sveriges skrädderiförbund.

Boëthius, B. & Heckscher, E. F. (eds.) (1938) *Svensk handelsstatistik 1637–1737: Samtida bearbetningar*. Stockholm: Thule.

Brottsbalken den 21 december 1962 samt översikt över ändringar i strafflagen under åren 1865–1960 jämte hänvisningar till Nytt Juridisk Arkiv. (1963) Stockholm: Nordstedts.

Embetz Skrå För Peruque Makare Embetet i Stockholm 1711, in *Meddelande från Nordiska museet*. 1897, pp. 95–98.

Gazette des atours de Marie-Antoinette. Garde-robe des atours de la reine. Gazette pour l'année 1782. Paris: Réunion des musées nationaux – Archives nationales. Imprimerie Escourbiac.

Komiterade för Stockholms skrädderiarbetareförening april 1882.

Kongl. Maj:ts nådige Förklaring, angående Lagens rätta förstånd uti 5. Cap. 10. §. Utsöknings-Balken, om Gäldenärs bysättning. Dat. 19. October. [1750].

Kongl. Maj:ts Förklaring, huru i anledning af Förordningen then 19 Maji 1756 med owilkorliga förskrifningar, hwilka Gardes- och andre Soldater samt Ryttare och Dragoner af the wäfwade och indelte Regementer och Båtsmän kunna utgifwa, bör förhållas. Dat. 12 October. [1757].

Kongl. Swea Håf-Rätts Bref, angående beswär i mål, rörande bysättning å then, hwilke hos Domaren sökt cessionem bonorum. Dat. 20 December. [1757].

Kongl. Swea Hofrätts Bref til derunder hörande domstolar om skyndesamt afgörande af cessions och concurs twister [1768].

Kongl. Maj:ts Nådige Förordning Angående Rymmande för Gäld. Gifwen i Stockholm i Råd-Cammaren then 11. December. 1766.

Kongl. Maj:ts nådiga Förklaring, om hwad wid Fallisementer samt Cessions- och Concours twister hädanefter bör i akt tagas. Dat. Then 8 Maji. [1767].

Kongl. Maj:ts förnyade Stadga, angående Afträdes- och Förmåns- samt Boskilnads och Urarfwa mål. Dat. then 26 Augusti. [1773].

Kongl. Maj:ts förnyade Stadga, angående Afträdes- och Förmåns- samt Boskilnads och Urarfwa mål. Dat. then 26 Augusti. [1773].

Konkurstillsynsutredningen (2000) *Ny konkurstillsyn. Betänkande. D. 1 Överväganden och förslag*. (SOU 2000:62) Stockholm: Fritzes offentliga publikationer.

Landshövdingarnas femårsberättelser. (1822–1855, Kungl. Maj:ts befallningshafvandes).

Lesage, A.-R. (1709) *Turcaret*.

Mercier, L.-S. (1782–1788) *Tableau de Paris*.

Molière. (1682a) *L'École des maris*.

Molière. (1682b) *Dom Juan*.

Molière. (1963) *Le bourgeois gentilhomme*. France: Larousse.

Molière. (2009) *The Miser (L'Avare)*.

Montesquieu. (1721) *The complete works of M. de Montesquieu*, 4 vols.

Nordisk familjebok. 38 vols. (1904–1926). Stockholm: Nordisk familjeboks förlag.

Nordisk familjebok. 23 vols (1923–1937). Stockholm: Nordisk familjeboks förlag.

Nordisk Retsencyklopaedi. 5 vol. (1878–1890). Kjøbenhavn: Gyldendal.

Prévost, N. (1792) *Grande dispute entre Marie-Antoinette et ses fournisseurs, traiteurs, tailleurs, marchandes de mode, &c.,&c.,&c., se voyant forçée à les payer, et la Nation ne voulant plus lui fournir des fonds.* Paris: Imprimerie de Feret.

Orsaken til närwarande talrika bankerotter och medlen att förekomma dem. Öfwerlämnad Patrioternas granskning av en Werldsmedborgare. (1799). Karlskrona: Kongl. M:ts Boktryckeri.

Preliminär textilteknisk ordlista. Maj 1957. (1957). Stockholm.

Rousseau, J. J. (1992). The collected writings of Rousseau, vol. 3, *Discourse on the origins of inequality (second discourse) polemics and political economy*, Masters, R. D. and Kelly, C. (eds.). Hanover: University Press of New England.

Smith, A. ([1762–63] 1981–1987). *The Glasgow edition of the works and correspondence of Adam Smith.* Vol. V: Lectures on Jurisprudence - Report of 1762–63. Oxford: Clarendon.

Smith, A. (1776 [1967]) *An inquiry into the nature and causes of the wealth of nations.* Oxford: Clarendon.

Sveriges Rikes Lag...(1884) *...med tillägg af de stadganden, som utkommit till den 20 december 1884 jämte bihang...* Stockholm: Norstedt.

Underdånigt Betänkande till Kongl. Maj:t, angående Kreditförhållandenes och Låneanstalternes ordnande avgivet den 8 april 1853 af särskildt i Nåder utsedde Comiterade. (1853). Stockholm: Nordstedt & Söner.

Äldre Swenska och ännu gällande lagar emot ocker i allmänhet, Spannmålspräjeri, Silfwersedlar, Agiotering, och det redbara Myntets Utpractiserande ur Riket. (1816) Stockholm: Elméns och Granbergs Tryckeri.

Unpublished literature and manuscripts

Allen, R. C. & Weisdorf, J. L. (2010) "Was there an 'industrious revolution' before the industrial revolution? An empirical exercise for England, c. 1300–1830", *Discussion Papers*, 10–14, Department of Economics, University of Copenhagen.

Coquery, N. (2006) "La boutique à Paris au XVIII siècle", *Habilitation à diriger des recherches*, Histoire, Université Paris I Panthéon Sorbonne.

Johansson Åbonde, A. (2010). "Drömmen om svenskt silke. Silkesodlingens historia i Sverige 1735–1920", *Licentiatavhandling*, Sveriges lantbruksuniversitet, Alnarp.

Lane, S. J. & Schary, M. A. (1989) "The macroeconomic component of business failures, 1956–1988", *Working paper*, School of Management, Boston University.

Larsson, P. (2016) "Marknadsföring i skråväsendets skugga. En undersökning av möbelannonser i Stockholms Dagblad 1840–1870", *B-uppsats*, Ekonomiskhistoriska institutionen, Uppsala universitet.

Mikhed, V. & Scholnick, B. (2014) "How do exogenous shocks cause bankruptcy? Balance sheet and income statement channels", *Working Paper*, 14–17, Federal Reserve Bank of Philadelphia.

Paulsson, S. (2004) "Ekonomiskt omyndig! Synen på kvinnlig egendom inom äktenskapet speglad i grosshandlarhustrurs försvar av sina hemgifter. Grosshandlarkonkurser i Stockholm 1781–1799", *C-uppsats*, Ekonomisk-historiska institutionen, Uppsala universitet.

Ruotsin peruukintekijät 1648–1810. I. Peruukintekijät, kampaajat, hiuskauppiaat ja hiustenkäsittelijät sekä puuteritehtailijat Tukholmassa 1648–1810 [Swedish wigmakers 1648–1810. I. Wigmakers, coiffeurs, hair traders and hair stylers and powder manufacturer in Stockholm 1648–1810], manuscript, by Kustaa, H. J. Vilkuna.

Schnabel, I. & Shin, H. S. (2001) "Foreshadowing LTCM: The Crisis of 1763", *Papers from Sonderforschungsbreich*, 504, London School of Economics.

Stadius, C. (2013) "En studie av stolstillverkningen i Stockholm 1720–1820", *C-uppsats*, Institutionen för kulturvård, Göteborgs universitet.

Literature

Adamson, R. (1966) *Järnavsättning och bruksfinansiering 1800–1860.* Diss. Göteborg: Göteborgs universitet.

Agge, I. (1934) "Några drag av det svenska konkursförfarandets utveckling", in Westman, K. G., Ekeberg, B., Forssner, T., Schlyter, K. Wedberg, B. & Strahl, I. (eds.), *Minnesskrift ägnad 1734 års lag av jurister i Sverige och Finland.* Stockholm: Svensk juristtidning, pp. 915–937.

Ahlberger, C. (1996) *Konsumtionsrevolutionen. Om det moderna konsumtionssamhällets framväxt 1750–1900.* Göteborg: Göteborgs universitet.

Albinsson Bruhner, G. (2004) *Att lyss till en sträng som brast.* Stockholm: Ratio.

Aldman, L.-A. (2008) *En merkantilistisk början. Stockholms textila import 1720–1738.* Diss. Uppsala: Uppsala universitet.

Alexander, J. (1891) *Konkursgesetze aller Länder der Erde. Mit vergleichender Übersicht.* Berlin: Puttkammer & Mühlbrecht.

Alm, G. (1993) *Carl Hårleman och den svenska rokokon.* Lund: Signum.

Amira von, K. (1913) *Grundriss des germanischen Rechts.* Strassburg: Verlag von Karl J. Trübner.

Andersson, B. (1988) *Göteborgs handlande borgerskap 1750–1805.* Göteborg: Göteborgs universitet.

Andersson, B., Fritz, M., & Olsson, K. (1996) *Göteborgs historia. Näringsliv och samhällsutveckling.* vol. 1. Stockholm: Nerenius & Santérus.

Andersson, T. & Sandberg, P. (2018) *Göteborgs historia.* Lund: Historiska media.

Andrén, E. (1973) *Snickare, schatullmakare och ebenister i Stockholm under skråtiden.* Stockholm: Nordiska museet.

Appel, C. & Skovgaard-Petersen, K. (2013) "The history of the book in the Nordic countries", in Suarez M. F. & Wuodhuysen H. R. (eds.), *The book. A global history.* Oxford: Oxford University Press, pp. 393–405.

Arrunada, B. (1996) "The economics of notaries" *European Journal of Law and Economics.* 3(1), pp. 5–37.

Axberger, H.-G. (2016) "Den svenska tryckfrihetens rättsliga utveckling", in Wennberg, B. & Örtenhed, K. (eds.) *Fritt ord 250 år. Tryckfrihet och offentlighet i Sverige och Finland – ett levande arv från 1766.* Stockholm: Sveriges Riksdag, pp. 241–288.

Balcaen, S. & Ooghe, H. (2006) "35 years of studies on business failure: an overview of the classic statistical methodologies and their related problems" *The British Accounting Review.* 38(1), pp. 63–93.

Balleisen, E. J. (2001) *Navigating failure. Bankruptcy and commercial society in Antebellum America.* Chapel Hill: University of North Carolina Press.

Barton, H. A. (1977) "Canton at Drottningholm: A model manufacturing community from the mid-eighteenth century" *Scandinavian Studies.* 49, pp. 81–98.

Barton, H. A. (1985) *Canton vid Drottningholm: ett mönstersamhälle för manufakturer från 1700-talet.* Spånga: Arena.

Barton, H. A. (1994) "Uppkomsten av manufakturerna i Canton vid Drottningholm: några tankar kring en feldatering", in Hallerdt, B. (ed.), *Sankt Eriks årsbok*. Stockholm: Samfundet S:t Erik, pp. 42–54.

Befolkningsutvecklingen under 250 år. Historisk statistik för Sverige (1999). Stockholm: Statistiska centralbyrån.

Beijer, A. (1981) *Drottningholms slottsteater på Lovisa Ulrikas och Gustaf III:s tid.* Stockholm: Liber Förlag.

Benson, J. & Shaw, G. (eds.) (1992) *The evolution of retail systems, c.1800–1914.* Leicester: Leicester University Press.

Berg, J. & Stavenow-Hidemark, E. (2001) "Stockholms Stads Sidenmanufactorie. En stor satsning med en kort historia" *RIG - Kulturhistorisk tidskrift*. 84, pp. 82–91.

Berg, M. (2005) *Luxury and pleasure in Britain.* Oxford: Oxford University Press.

Bergfalk, P. E. (1859) *Bidrag till de under de sista hundrade åren inträffade handelskrisers historia.* Uppsala: Edquist & K.

Berglund, M. (2009) *Massans röst. Upplopp och gatubråk i Stockholm 1719–1848.* Diss. Stockholm: Stockholms universitet.

Bertucci, P. (2013) "Enlightened secrets: silk, intelligent travel, and industrial espionage in eighteenth-century France" *Technology and Culture*. 54, pp. 820–852.

Björklund, A. (2010) *Historical urban agriculture. Food production and access to land in Swedish towns before 1900.* Diss. Stockholm: Stockholms unversitet.

Bladh, C. (1991) *Månglerskor Att sälja från korg och bod i Stockholm 1819–1846.* Diss. Göteborg: Göteborgs universitet.

Bladh, C. (ed.) (2008) *Rodderskor på Stockholms vatten.* Stockholm: Stockholmia förlag.

Blondé, B. & van Damme, I. (2010) "Retail growth and consumer changes in a declining urban economy" *Economic History Review*. 63(3), pp. 638–663.

Box, M. (2005) *New venture, survival, growth.* Diss. Stockholm: Stockholms universitet.

Box, M. (2008) "The death of firms: exploring the effects of environment and birth cohort on firm survival in Sweden" *Small Business Economics*. 3, pp. 379–393.

Box, M., Gratzer, K., & Lin, X. (2016) "Konkurs och konjunktur i Sverige 1830–2010" *Insolvensrättslig tidskrift*. 1, pp. 20–36.

Boëthius, B. (1943) *Magistraten och borgerskapet i Stockholm 1719–1815.* Stockholm: Norstedt.

Braudel, F. (1981) *Civilization and capitalism: 15th–18th century.* vol. 1. London: Collins.

Braudel, F. (1986) *Marknadens spel. Civilisationer och kapitalism 1400–1800* (Les Jeux de L'Échange). vol. 2. Stockholm: Gidlunds.

Braudel, F. (1997) *Medelhavet och medelhavsvärlden på Filip II:s tid.* (La Méditerranée et le monde méditerranéen à l'époque de Philippe II). Furulund: Alhambra.

Bremer-David, C. (ed.) (2011) *Paris. Life and luxury in the eighteenth century.* [exhibition, J. Paul Getty museum, Getty Center, Los Angeles, Apr. 26–Aug. 7, 2011; Museum of fine arts, Houston, Sept. 18–Dec. 10, 2011]. Los Angeles: J. Paul Getty Museum.

Brennan, T. (2013) "Debt and default in 18th-century champagne", in Safley, T. M. (ed.), *The history of bankruptcy. Economic, social and cultural implications in early modern Europe.* New York: Routledge, pp. 34–51.

Brewer, J. & Porter, R. (ed.) (1993) *Consumption and the world of goods.* London: Routledge.

Bring, S. E. (1934) "Strinnholms båda arbeten om Magnus Stenbock" *Nordisk tidskrift för bok- och biblioteksväsen*, 21, pp. 40–44.

Bring, S. E. & Kulling, E. (1943) *Svenska Boktryckareföreningen 50 år. Kort historik över Boktryckeri-Societeten och Svenska Boktryckareföreningen utgiven med anledning av föreningens femtioårs jubileum.* Stockholm: Svenska Boktryckareföreningen.

Broberg, O. (2006) *Konsten att skapa pengar. Aktiebolagens genombrott och finansiell modernisering kring sekelskiftet 1900.* Diss, Göteborg: Göteborgs universitet.

Brolin, P.-E. (1953) *Hattar och mössor i borgarståndet 1760–1766.* Diss. Göteborg: Göteborgs högskola.

Broomé, G. (1888) *Studier i konkursrätt särskildt med afseende på svensk rätt.* Diss. Lund: Gleerup.

Brown, D. (2000) "'Persons of infamous character' or 'an honest, industrious and useful description of people'? The textile pedlars of Alstonfield and the role of peddling in industrialization" *Textile History.* 31(1), pp. 1–26.

Browning, A. H. (2019) *The panic of 1819. The first great depression.* Columbia: University of Missouri Press.

Bull, I. (2011) "Industriousness and development of the school-system in the eighteenth century: the experience of Norwegian cities" *History of Education.* 40(4), pp. 425–446.

Burkard, S. (2010) *Mémoires de la Baronne d'Oberkirch sur la cour de Louis XVI et la société française avant 1789.* Paris: Mercure de France.

Börjeson, D. Hj. T. (1932) *Stockholms segelsjöfart 1732–1932.* Stockholm: Sjokaptenssocieteten.

Campagnol, I. (2014) *Forbidden fashions. Invisible luxuries in early Venetian convents.* Lubbock: Texas Tech University Press.

Carboni, M. & Massimo, F. (2013) "Learning from others' failures: the rise of the Monte di pieta in early modern Bologna", in Safley, T. M. (ed.), *The history of bankruptcy. Economic, social and cultural implications in early modern Europe.* New York: Routledge, pp. 108–125.

Cavallo, S. & Warner, L. (eds.) (1999) *Widowhood in medieval and early modern Europe.* London: Longman.

Cederblom, G. (1927–1929) *Pehr Hilleström som kulturskildrare.* 2 vols. Stockholm: Nordiska museet.

Chandler, A. D. (1977) *The visible hand. The managerial revolution in American business.* Cambridge: Belknap.

Chaudhury, S. & Morineau, M. (eds.) (1999) *Merchants, companies and trade. Europe and Asia in the early modern era.* Cambridge: Cambridge University Press.

Ciszuk, M. (2012) *Silk weaving in Sweden during the 19th century. Textiles and texts – an evaluation of the source material.* Göteborg: Chalmers University of Technology.

Ciszuk, M. (2013) "Sidenvävning under 1800-talet. Teknikutveckling och marknad – sidentyger i Enebergs samling", in Nyberg, K. & Lundqvist, P. (eds.), *Dolda innovationer. Textila produkter och ny teknik under 1800-talet.* Stockholm: Kulturhistoriska bokförlaget, pp. 83–120.

Clark, G. & Van Der Werf, Y. (1998) "Work in progress? The industrious revolution" *The Journal of Economic History.* 58(3), pp. 830–843.

Coffin, J. G. (1996) *The politics of women's work. The Paris garment trades, 1750–1915,* Princeton: Princeton University Press.

Cole, C. W. (1939) *Colbert and a century of French mercantilism.* 2 vols. New York: Columbia University Press.

Cole, C. W. (1943) *French mercantilism, 1683–1700.* New York: Columbia University Press.

Coleman, P. J. (1974) *Debtors and creditors in America: insolvency, imprisonment for debt, and bankruptcy, 1607–1900.* Madison: State Historical Society of Wisconsin.

Coquery, N. (2004) "The language of success. Marketing and distributing semi-luxury goods in eighteenth-century Paris" *Journal of Design History.* 17(1), pp. 71–89.

Coquery, N. (2009) "The Semi-luxury market, shopkeepers and social diffusion: Marketing *Chinoiseries* in eighteenth-century Paris", in Blondé, B., Coquery, N., Stobart, J., & Van Damme, I. (eds.), *Fashioning old and new: changing consumer patterns in western Europe (1650–1900).* Turnhout: Brepols, pp. 121–131.

Coquery, N. (2013) "Credit, trust and risk. Shopkeepers' bankruptcies in 18th-century", in Safley, T. M. (ed.) *The history of bankruptcy. Economic, social and cultural implications in early modern Europe.* New York: Routledge, pp. 52–71.

Cordes, A. & Schulte Beerbühl, M. (eds.) (2016) *Dealing with economic failure. Between norm and practice* (15th to 21st century). Frankfurt am Main: Peter Lang.

Crowston, C. H. (2001) *Fabricating women. The seamstresses of old regime France, 1675–1791.* Durham: Duke University Press.

Crowston, C. H. (2009) "The Queen and her 'minister of fashion': gender, credit and politics in pre-revolutionary France", in McNeil, P. (ed.) *Fashion. Critical and primary sources. The eighteenth century,* vol. 2. Oxford & New York: Berg, pp. 192–215.

Crowston, C. H. (2013) *Credit, fashion, sex. Economies of regard in old regime France.* Durham & London: Duke University Press.

Cruz, L. & Mokyr, J. (eds.) (2010) *The birth of modern Europe. Culture and economy, 1400–1800. Essays in honor of Jan de Vries.* Leiden: Brill.

Cumming, V., Cunnington, C. W., & Cunnington, P. E. (2017) *The dictionary of fashion history.* London: Bloomsbury Academic.

Curtis, L. P. (1980) "Incumbered wealth: landed indebtedness in post-famine Ireland" *The American Historical Review.* 85(2), pp. 332–367.

Danielsson, I.-M. (1998) *Den bildade smaken. Målade dekorationer hos borgerskapet i frihetstidens Stockholm.* Diss. Stockholm: Stockholms universitet.

Das Gupta, A. (2001) *The world of the Indian ocean merchant 1500–1800.* Oxford: Oxford University Press.

Dehing, P. & 't Hart, M. (1997) "Linking the fortunes: currency and banking, 1550–1800", in T Hart, M, Jonker, J., & Zanden, J, Luiten van. (eds.), *A financial history of the Netherlands.* Cambridge: Cambridge University Press.

Delpierre, M. (1996) *Se vêtir au XVIIIe siècle.* Paris: Adam Biro.

Dermineur, E. M. (2018) "Peer-to-peer lending in pre-industrial France" *Financial History Review.* 26(3), pp. 359–388.

Di Martino, P. (2005) "Approaching disaster: personal bankruptcy legislation in Italy and England, c. 1880–1939" *Business History.* 47(1), pp. 23–43.

Dreutzer, O. M. (1844) *Handbok för handlande och sjöfarande af juridiskt, administrativt och ekonomiskt innehåll.* Örebro: N.N.

Duffy, I. P. H. (1980) "English bankrupts, 1571–1861" *The American Journal of Legal History.* 24(4), pp. 283–305.

Duffy, I. P. H. (1985) *Bankruptcy and insolvency in London during the industrial revolution.* London and New York: Garland/Routledge.

Edvinsson, R. (2005) *Growth, accumulation, crisis. With new macroeconomic data for Sweden 1800–2000.* Stockholm: Almqvist & Wiksell International.

Edvinsson, R., Jacobson, T., & Waldenström, D. (eds.) (2010) *Exchange rates, prices, and wages, 1277–2008.* Stockholm: Ekerlid.

Edvinsson, R., Jacobson, T., & Waldenström, D. (eds.) (2014) *House prices, stock returns, national accounts, and the Riksbank balance sheet, 1620–2012.* Stockholm: Ekerlids förlag.

Edwards, C. D. (1996) *Eighteenth-century furniture.* Manchester: Manchester University Press.

Edwards, C. D. (2005) *Turning houses into homes. A history of the retailing and consumption of domestic furnishings.* London: Routledge.

Eggeby, E. & Nyberg, K. (2002) "Stad i stagnation 1720–1850", in Nilsson, L. (ed.) *Staden på vattnet.* vol. 1. Stockholm: Stockholmia förlag, pp. 185–276.

Ekegård, E. (1924) *Studier i svensk handelspolitik under den tidigare frihetstiden.* Diss. Uppsala: Uppsala universitet.

Engellau-Gullander, C. (2001) "Jean Eric Rehn och nyttokonsten på 1700-talet. En historiografisk studie" *Konsthistorisk tidskrift.* 70, pp. 171–188.

Epstein, S. (1991) *Wage labor & guilds in medieval Europe.* Chapel Hill: University of North Carolina Press.

Erickson, A. L. (1995[1993]) *Women and Property in Early Modern England.* London, New York: Routledge.

Erikson, M. (2018) *Krediter i lust och nöd. Skattebönder i Torstuna härad, Västmanlands län, 1770–1870.* Diss. Uppsala: Acta Universitatis Upsaliensis.

Erler, A. (1990) "Schuldhaft", in Erler, A. & Kauffman, E. (eds.) *Handwörterbuch zur deutschen Rechtsgeschichte*, vol. 2. Berlin: E. Schmidt, pp. 1511–1513.

Everett, J. & Watson, J. (1998) "Small business failure and external risk factors" *Small Business Economics.* 11(4), pp. 371–390.

Fairchilds, C. (1993) "The production and marketing of populuxe goods in eighteenth-century Paris", in Brewer, J. & Porter, R. (eds.), *Consumption and the world of goods,* London: Routledge, pp. 228–248.

Falck, P. (2008a) "Fredrik I:s tid. Senbarock 1720-talet till 1750-talet", in Nyström, B. & Eklund Nyström, S. (eds.), *Svenska möbler under femhundra år.* Stockholm: Natur & kultur, pp. 62–87.

Falck, P. (2008b) "Rokoko. 1750-talet till 1770-talets mitt", in Nyström, B. & Eklund Nyström, S. (eds.), *Svenska möbler under femhundra år.* Stockholm: Natur & kultur, pp. 88–127.

Falck, P. & Westberg, P. (1996) *"Möbelsnickare"*, in *Hantverk i Sverige. Om bagare, kopparslagare, vagnmakare och 286 andra hantverksyrken,* Nyström, B., Biörnstad, A., & Bursell, B. (eds.), Stockholm: LT & Nordiska museet, pp. 88–127 and 236–244.

Falk, U. & Kling, C. (2016) "The regulatory concept of compulsory composition in the German bankruptcy act", in Cordes, A. & Schulte Beerbühl, M. (eds.) (2016) *Dealing with economic failure. Between norm and practice* (15th to 21st century). Frankfurt am Main: Peter Lang, pp. 215–241.

Fine, B. & Leopold, E. (1993) *The world of consumption.* London: Routledge.

Finn, M. C. (2003) *The character of credit. Personal debt in English culture, 1740–1914.* Cambridge: Cambridge University Press.

Fischer, P. (2013) "Bankruptcy in early-modern German territories", in Safley, T. M. (ed.), *The history of bankruptcy. Economic, social and cultural implications in early modern Europe.* New York: Routledge, pp. 171–184.

Fleet, K. (1999) *European and Islamic trade in the early Ottoman state. The merchants of Genoa and Turkey.* Cambridge: Cambridge University Press.

Fontaine, L. (2001) "Antonio and Shylock: credit and trust in France, c. 1680–c. 1780" *The Economic History Review.* 54(1), pp. 39–57.

Fontaine, L. (2014[2008]) *The moral economy. Poverty, credit, and trust in early modern Europe.* Cambridge: Cambridge University Press.

Forsstrand, C. (1918) *Storborgare och stadsmajorer. Minnen och anteckningar från Gustaf III:s Stockholm.* Stockholm: Geber.

Fridenson, P. (2004) "Business failure and the agenda of business history" *Enterprise & Society.* 5(4), pp. 562–582.

Fryde, E. (1996) "The bankruptcy of the Scali of Florence in England 1326–1328", in Britnell, R. & Hatcher, J. (eds.), *Progress and problems in medieval England. Essays in honour of Edward Miller.* Cambridge: Cambridge University Press, pp. 107–120.

Fällström, A. M. (1974) *"Näringsliv och levnadsvillkor i Göteborg 1800–840"*, in Fallstrom, A. M. (ed.), *Konjunktur och kriminalitet. Studier i Göteborgs sociala historia 1800–840.* Goteborg: Goteborgs universitet, pp. 5–0 and 37–2.

Gelderblom, O. (ed.) (2013) *Cities of commerce. The institutional foundations of international trade in the Low Countries, 1250–1650.* Princeton: Princeton University Press.

Grandell, A. (1944) *Äldre redovisningsformer i Finland. En undersökning av den företagsekonomiska redovisningens utveckling i Finland intill 1800-talets slut.* Diss. Helsingfors: Helsingfors universitet.

Grassby, R. (1995) *The business community of seventeenth-century England.* Cambridge: Cambridge University Press.

Gratzer, K. (1995) "Vie et mort des entreprises à Stockholm entre 1899 et 1990. Vers une démographie des entreprises", in Moss, M. & Jobert, P. (eds.), *Naissance et mort des entreprises en Europe XIXe - XXe siecles.* Dijon: Éditions de l'Université de Dijon, pp. 157–178.

Gratzer, K. (1999) "Konkursens orsaker", in Gratzer, K. & Sjögren, H. (eds.), *Konkursinstitutets betydelse i svensk ekonomi.* Hedemora: Gidlund, pp. 134–159.

Gratzer, K. (2008a) "Default and imprisonment for debt in Sweden: from the lost chances of a ruined life to the lost capital of a bankrupt company", in Gratzer, K & Stiefel, D. (eds.), *History of insolvency and bankruptcy from an international perspective.* Huddinge: Södertörns högskola, pp. 15–59.

Gratzer, K. (2008b) "Introduction", in Gratzer, K. & Stiefel, D. (eds.), *History of insolvency and bankruptcy from an international perspective.* Huddinge: Södertörns högskola, pp. 5–14.

Gratzer, K. & Box, M. (2002) "Causes of selection amongst Swedish firms: A contribution to the development of a business demography" *Scandinavian Economic History Review.* 50(1), pp. 68–84.

Gratzer, K. & Sjögren, H. (eds.) (1999) *Konkursinstitutets betydelse i svensk ekonomi.* Hedemora: Gidlund.

Gratzer, K. & Stiefel, D. (eds.) (2008) *History of insolvency and bankruptcy from an international perspective.* Huddinge: Södertörns högskola.

Greif, A. (2006) *Institutions and the path to the modern economy. Lessons from medieval trade.* New York: Cambridge University Press.

Grimm, J. (1881) *Deutsche Rechtsalterthümer.* Göttingen: Dieterische Reichsbuchhandlung.

Gross, K., Newman, M. S., & Campbell, D. (1996) "Ladies in red: learning from America's first female bankrupts" *The American Journal of Legal History.* 40(1), pp. 1–40.

Gråbacke, C. (2015) *Kläder, shopping och flärd: modebranschen i Stockholm 1945–2010*. Stockholm: Stockholmia förlag.

Gårdlund, T. (1947) *Svensk industrifinansiering under genombrottsskedet 1830–1913*. Stockholm: Svenska bankföreningen.

Göransson, A. (1990) "Kön, släkt och ägande. Borgerliga maktstrategier 1800–1850" *Historisk tidskrift*. 4, pp. 525–543.

Göransson, J. (1937) *Aftonbladet som politisk tidning 1830–1835*. Diss. Uppsala: Wretman.

Gösche, A. (1985) Insolvenzen und wirtschaftlicher wandel. Eine wirtschaftsgeschichtliche analyse der konkurse und vergleiche im Siegerland 1951–1980. Wiesbaden: Steiner.

Hagberg, A. & Nyberg, K. (2017a) "Appendix 1. Måleriets och bildkonstens sociala förhållanden 1720–1850", in in Nyberg, K. (ed.) *Ekonomisk kulturhistoria. Bildkonst, konsthantverk och scenkonst 1720–1850*. Stockholm: Kulturhistoriska bokförlaget, pp. 183–187.

Hagberg, A. & Nyberg, K. (2017b) "Appendix 2. Måleri och bildkonst i Stockholm 1720–1850", in Nyberg, K. (ed.) *Ekonomisk kulturhistoria. Bildkonst, konsthantverk och scenkonst 1720–1850*. Stockholm: Kulturhistoriska bokförlaget, pp. 189–214.

Hammar, B. & Rasmussen, P. (2001) *Kvinnligt mode under två sekel*. Lund: Signum.

Haneson, V. & Rencke, K. (1923) *Bohusfisket*. Göteborg: Göteborgs Litografiska.

Hansen, B. (1998) "Commercial associations and the creation of a national economy. The demand for federal bankruptcy law" *Business History Review*. 72(1), pp. 86–113.

Hasselberg, Y. (1998) *Den sociala ekonomin. Familjen Clason och Furudals bruk 1804–1856*. Diss. Uppsala: Uppsala universitet.

Hayami, A. (2015) *Japan's industrious revolution. Economic and social transformations in the early modern period*. Tokyo: Springer.

Hayen, Mats (2007) *Stadens puls. En tidsgeografisk studie av hushåll och vardagsliv i Stockholm, 1760–1830*. Diss. Stockholm: Stockholms universitet.

Hayen, M. (2017) "Ett hav av omöjligheter. Konkurser för personal vid Kungliga Teatern, 1773–1792", in Nyberg, K. (ed.) *Ekonomisk kulturhistoria. Bildkonst, konsthantverk och scenkonst 1720–1850*. Stockholm: Kulturhistoriska bokförlaget, pp. 141–155.

Hayen, M. & Nyberg, K. (2017) "Konkursinstitutet i Stockholm 1720–1850", in Nyberg, K. (ed.), *Ekonomisk kulturhistoria. Bildkonst, konsthantverk och scenkonst 1720–1850*. Stockholm: Kulturhistoriska bokförlaget, pp. 37–48.

Heckscher, E. F. (1918) *Kontinentalsystemet. Den stora handelsspärrningen för hundra år sedan*. Stockholm: Norstedts.

Heckscher, E. F. (1935–1949) *Sveriges ekonomiska historia från Gustav Vasa*. Stockholm: Bonnier.

Heckscher, E. F. (1937) "De svenska manufakturerna under 1700-talet" *Ekonomisk Tidskrift*. 39, pp. 153–221.

Heckscher, E. F. (1994[1935]) *Mercantilism*. 2 vols. London: Routledge.

Heidner, J. (1995) "Carl Gustaf Tessin: en samlare och konstförmedlare", in von Proschwitz, Gunnar (ed.), *Carl Gustaf Tessin: kulturpersonen och privatmannen: 1695–1770*. Stockholm: Atlantis, pp. 21–40.

Hellman, M. (1999) "Furniture, sociability, and the work of leisure in eighteenth century France" *Eighteenth-Century Studies*, 32(4), pp. 415–445.

Henderson, W. O. (1985) *Manufactories in Germany.* Frankfurt am Main: Peter Lang.

Heurlin, F., Millqvist, V. & Rubenson, O. (1906) *Svenskt biografiskt handlexikon.* Stockholm: Bonnier.

Hildebrand, B. E. (1874–1875) *Sveriges och svenska konungahusets minnespenningar, praktmynt och belöningsmedaljer.* 2 vols. Stockholm: Kongl. vitterhets historie och antiqvitets akademiens förlag.

Hinners, L. (2012) *De fransöske handtwerkarne vid Stockholms slott 1693–1713. Yrkesroller, organisation, arbetsprocesser.* Diss. Stockholm: Stockholms universitet.

Hjern, H. (1943) *Beklädnadsarbetare. Historik över Svenska beklädnadsarbetareförbundets avdelning 8:s 60-åriga fackliga arbete,* 1882–1942. Göteborg: Framåt.

Hodacs, H. (2016) *Silk and tea in the north. Scandinavian trade and the market for Asian goods in eighteenth-century Europe.* Basingstoke: Palgrave Macmillan.

Hoffman, P. T., Gilles, P.-V., & Rosenthal, J.-L. (2000) *Priceless markets. The political economy of credit in Paris, 1660–1870.* Chicago: University of Chicago Press.

Hofmeester, K., & Grewe, B.-S. (eds.) (2016) *Luxury in global perspective. Objects and practices, 1600–2000.* New York: Cambridge University.

Hoppit, J. (1987) *Risk and failure in English business, 1700–1800.* Cambridge: Cambridge University Press.

Hunt, E. S. & Murray, J. M. (1999) *A history of business in medieval Europe, 1200–1550.* Cambridge: Press Syndicate of the University of Cambridge.

Hutchison, R. (2012) *In the doorway to development. An enquiry into market oriented structural changes in Norway ca. 1750–1830.* Leiden: Brill.

Häberlein, M. (2013) "Merchants' bankruptcies, economic development and social relations in German cities during the long 16th century", in Safley, T. M. (ed.) *The history of bankruptcy. Economic, social and cultural implications in early modern Europe.* New York: Routledge, pp. 19–33.

Högberg, S. (1969) *Utrikeshandel och sjöfart på 1700-talet. Stapelvaror i svensk export och import 1738–1808.* Stockholm: Bonniers.

Ilmakunnas, J. (2012) *Ett ståndsmässigt liv. Familjen von Fersens livsstil på 1700-talet.* Helsingfors: Svenska litteratursällskapet i Finland.

Ilmakunnas, Johanna (2015) "The luxury shopping experience of the Swedish aristocracy in eighteenth-century Paris", in Simonton, D., Kaartinen, M., & Montenach, A. (eds.), *Luxury and gender in European towns, 1700–1914.* New York: Routledge, pp. 115–131.

Ilmakunnas, J. (2016) *Joutilaat ja ahkerat. Kirjoituksia 1700-luvun Euroopasta.* Helsinki: Siltala.

Ilmakunnas, J. (2017) "French fashions: aspects of elite lifestyle in eighteenth-century Sweden", in Ilmakunnas, J. & Stobart, J. (eds.), *A taste for luxury in early modern Europe. Display, acquisition and boundaries.* London: Bloomsbury Academic, pp. 243–263.

Ilmakunnas, J., Rahikainen, M., & Vainio-Korhonen, K. (eds.) (2018) *Early professional women in northern Europe, c. 1650–1850.* London: Routledge.

Inger, G. (1997) *Svensk rättshistoria.* Malmö: Liber.

Jansson, M. (2017) *Making metal making. Circulation and workshop practices in the Swedish metal trades, 1730–1775.* Diss. Uppsala: Acta Universitatis Upsaliensis.

Jenkins, D. T. (ed.) (2003) *The Cambridge history of western textiles.* vol. 1, Cambridge: Cambridge University Press.

Jones, J. (2004) *Sexing la mode. Gender, fashion and commercial culture in old regime France.* Oxford and New York: Berg.

Jones, J. W. (1979) *The foundations of English bankruptcy. Statutes and commissions in the early modern period.* Philadelphia: American Philosophical Society.

Jones, R. M. (2006) *The apparel industry.* 2nd ed. Oxford: Blackwell.

Jonker, J. (1996) *Merchants, bankers, middlemen. The Amsterdam money market during the first half of the 19th century.* Amsterdam: NEHA.

Joseph, J. (2013) "La littérature et la presse au XVII et au XVIII siècle" *Romanistische Zeitschrift für Literaturgeschichte,* 37, pp. 81–106.

Jägerskiöld, Olof (1945) *Lovisa Ulrika.* Stockholm: Wahlström & Widstrand.

Jörberg, L. (1972) *A history of prices in Sweden 1732–1914.* 2. vols. Lund: Gleerup.

Kawamura, Y. (2005) *Fashion-ology. An introduction to fashion studies.* Oxford: Berg.

Kermode, J. (1998) *Medieval merchants. York, Beverley and Hull in the latter middle ages.* Cambridge: Cambridge University Press.

Kilborn, J. (2009) *Fartyg i Europas periferi under den industriella revolutionen. Den svenska utrikeshandelsflottan 1795–1845.* Göteborg: Göteborgs universitet.

Kindleberger, C. P. (1996[1978]) *Manias, panics and crashes. A history of financial crises.* London: Macmillan.

Kirby, D. (1990) *Northern Europe in the early modern period. The Baltic world 1492–1772.* London: Longman.

Kirby, D. (1995) *The Baltic world 1772–1993. Europe's northern periphery in an age of change.* London: Longman.

Kirkham, P. (1988) "The London furniture trade 1700–1870" *The Journal of the Furniture History Society,* 24, pp. 1–219.

Kjellberg, S. T. (1943) *Ull och ylle. Bidrag till den svenska yllemanufakturens historia.* Diss. Lund: Lunds universitet.

Klemming, G. E. & Nordin, J. G. (1883) *Svensk boktryckeri-historia 1483–1883.* Stockholm: Norstedt.

Kunstreich, J. (2016) "Bankruptcy laws as *Standortpolitik* – The case of Hamburg 1850 to 1870", in Cordes, A. & Schulte Beerbühl, M. (eds.), *Dealing with economic failure. Between norm and practice (15th to 21st century).* Frankfurt am Main: Peter Lang, pp. 193–214.

Kwass, M. (2006) "Big hair. A wig history of consumption in eighteenth-century France" *The American Historical Review.* 111(3), pp. 631–659.

Lagerquist, M. (1981) *Den yrkesmässiga möbelhandeln i Sverige intill år 1780. Studier i rokokotidens möbelhantverk och möbeldistribution.* Stockholm: Nordiska museet.

Laine, M. (1998) *"En Minerva för vår Nord". Lovisa Ulrika som samlare, uppdragsgivare och byggherre.* Diss. Uppsala: Uppsala universitet.

Lantmanson, I. S. (1866) *Om Concursbo, dess begrepp, uppkomst och omfattning.* Diss. Lund: Berlingska boktryckeriet.

Lausing, G. (1959) "Konjunkturen", in Beckerath, E. von (ed.), *Handwörterbuch der Sozialwissenschaften.* vol. 6. Göttingen: Vandenhoeck & Ruprecht, pp. 133–141.

Ledbetter, K. (2007) *Tennyson and Victorian periodicals – commodities in context (the nineteenth century).* Aldershot: Ashgate Publishing Limited.

Lee, J. (1999) "Trade and economy in preindustrial east Asia, c. 1500–c. 1800: East Asia in the age of global integration" *The Journal of Asian Studies.* 58(1), pp. 2–26.

Leijonhufvud, Madeleine. (1991) *Om brott mot borgenärer m.m.* Stockholm: Norstedts juridikförlag.

Leijonhufvud, Sigrid. (1915) *Carl Gustaf Tessins dagbok 1748–1752*. Stockholm: Norstedt.

Lemire, B. (1991) *Fashion's favourite. The cotton trade and the consumer in Britain, 1660–1800*. Oxford: Oxford University Press.

Lemire, B. (1997) *Dress, culture and commerce. The English clothing trade before the factory, 1660–1800*. Basingstoke: Macmillan.

Lemire, B. (1998) "Petty pawns and informal lending. Gender and the transformation of small-scale credit in England, circa 1600–1800", in Mathias, P., Bruland, K., & O'Brien, P. K. (eds.), *From family firms to corporate capitalism. Essays in business and industrial history in honour of Peter Mathias*. Oxford: Clarendon Press, pp. 112–138.

Lemire, B. (2012[2005]). *The business of everyday life. Gender, practice and social politics in England, c.1600–1900*. Manchester: Manchester University Press.

Lemire, B. (ed.) (2010) *The force of fashion in politics and society. Global perspectives from early modern to contemporary times*. Farnham: Ashgate.

Lemire, B. (2018) *Global trade and the transformation of consumer cultures. The material world remade, c. 1500–1820*. Cambridge: Cambridge University Press.

Lesger, C. & Noordegraaf, L. (eds.) (1995) *Entrepreneurs and entrepreneurship in early modern times. Merchants and industrialists within the orbit of the Dutch staple market*. Den Haag: Stichting Hollandse Historische Reeks.

Levy, A. & Barniv, R. (1987) "Macroeconomic aspects of firm bankruptcy analysis" *Journal of Macroeconomics*. 9(3), pp. 407–415.

Lilja, K. (2010) "The revolution of a market for household savers, 1820–1910", in Ögren, A. (ed.), *The Swedish financial revolution*. New York: Palgrave Macmillan, pp. 41–63.

Lilja, S. (1996) *Städernas folkmängd och tillväxt. Sverige (med Finland) ca 1570-tal till 1810-tal*. Stockholm: Stads-och kommunhistoriska institutet.

Lind, I. (1923) *Göteborgs handel och sjöfart 1637–1920. Historisk-statistisk översikt.* Göteborg: Skrifter utgivna till Göteborgs stads trehundraårsjubileum.

Lindberg, E. (2001) *Borgerskap och burskap. Om näringsprivilegier och borgerskapets institutioner i Stockholm 1820–1846*. Diss. Uppsala: Uppsala universitet.

Lindblom, A. (1924) *Jacques-Philippe Bouchardon och de franska bildhuggarna vid Stockholms slott under rokokotiden*. Uppsala: Almqvist & Wiksell.

Lindgren, H. (2010) "The Evolution of Secondary Financial Markets, 1820–1920", in Ögren, A. (ed.), *The Swedish financial revolution*. New York: Palgrave Macmillan, pp. 95–112.

Lindgren, H. (forthcoming) "Over-indebtedness in a pre-industrial society – or not? Debt accumulation and evolving debt structures in 19th century Sweden", for publication in *Scandinavian Economic History Review*.

Lindström, H. (1923) *Näringsfrihetens utveckling i Sverige 1809–1836*. Diss. Göteborg: Göteborgs Högskola.

Lindström, H. (1929) *Näringsfrihetsfrågan i Sverige 1837–1864*. Göteborg: Elanders Boktryckeri Aktiebolag.

MacLeod, J. (2013) "The reception of the Actio Pauliana in Scots law", in Safley, T. M. (ed.), *The history of bankruptcy. Economic, social and cultural implications in early modern Europe*. New York: Routledge, pp. 200–222.

Magnusson, L., & Nyberg, K. (1995) "Konsumtion och industrialisering i Sverige 1820–1914. Ett ekonomisk-historiskt forskningsprogram", *Uppsala papers in economic history, Research Reports*, nr. 38. Uppsala: Uppsala universitet, pp. 1–31.

Malmborg, G. (1927) "Jean Eric Rehns första verksamhetsår vid manufakturkontoret" *Fataburen*. 3, pp. 105–118.

Mann. B. H. (2002) *Republic of debtors. Bankruptcy in the age of American independence.* Cambridge: Harvard University Press.

Markham, L. V. (1995) *Victorian insolvency. Bankruptcy, imprisonment for debt, and company winding-up in nineteenth century.* Oxford: Clarendon.

Markovits, C. (2000) *The global world of Indian merchants, 1750–1947. Traders of Sind from Bukhara to Panama.* New York: Cambridge University Press.

Marriner, S. (1980) "English bankruptcy records and statistics before 1850" *Economic History Review*. 33(3), pp. 351–366.

Marsh, B. (2020) *Unravelled dreams. Silk and the Atlantic World, 1500–1840.* Cambridge: Cambridge University Press.

Martinius, B. (2008) "Gustaviansk tid. Nyklassicism. 1770-talet till 1810-talet", in Nyström, B. & Eklund Nyström, S. (eds.), *Svenska möbler under femhundra år.* Stockholm: Natur & kultur, pp. 128–157.

Mathias, P. (2000) "Risk, credit, kinship in early modern enterprise", in McCusker, J. J, & Morgan, K. (eds.), *The early modern Atlantic economy.* Cambridge: Cambridge University Press, pp. 15–35.

Matović, M. R. (1984) *Stockholmsäktenskap. Familjebildning och partnerval i Stockholm 1850–1890.* Diss. Stockholm: Stockholms universitet.

Matthee, R. P. (1999) *The politics of trade in Safavid Iran. Silk for silver, 1600–1730.* Cambridge: Cambridge University Press.

McKendrick, N., Brewer, J., & Plumb, J. H. (1983) *The birth of a consumer society. The commercialization of eighteenth-century England.* London: Hutchinson.

McNeil, P. & Riello, G. (2016) *Luxury. A rich history.* Oxford: Oxford University Press.

Meerkerk, E. Van Nederveen (2008) "Couples cooperating? Dutch textile workers, family labour and the industrious revolution', c. 1600–1800", *Continuity and Change.* 23, pp. 237–266.

Molà, L. (2000) *The silk industry of Renaissance Venice.* Baltimore: Johns Hopkins University Press.

Mortier, du Bianca M. (ed.) (2016) *Costume and fashion.* Amsterdam: Rijksmuseum.

Moselius, C. D. (1923) *Louis Masreliez. Med en inledning om Adrien och Jean Baptiste Masreliez' verksamhet på Stockholms slott.* Diss. Stockholm: Stockholms högskola.

Mosley, J. (2013) "The technologies of print", in Suarez, M. F., & Wuodhuysen, H. R. (eds.), *The book. A global history.* Oxford: Oxford University Press, pp. 130–153.

Mui, H.-C. & Mui, L. H. (1989) *Shops and shopkeeping in eighteenth-century England.* Québec: McGill-Queen's University Press.

Muldrew, C. (1998) *The economy of obligation. The culture of credit and social relationships in early modern England.* New York: St. Martin's Press.

Muldrew, C. (2011) *Food, energy and the creation of industriousness. Work and material culture in agrarian England, 1550–1780.* Cambridge: Cambridge University Press.

Murhem, S. (2016) "Advertising in a regulated economy. Swedish advertisments 1760–1800" *Journal of Historical Research in Marketing.* 8(4), pp. 484–506.

Müller, L. (2018) *Sveriges första globala århundrade. En 1700-talshistoria.* Stockholm: Dialogos.

Müller, L. Rydén, G. & Weiss, H. (2010) *Global historia från periferin. Norden 1600–1850.* Lund: Studentlitteratur.

Möller, B. (1945) "Deleen, Carl Erik", in Hildebrand, B. (ed.), *Svenskt biografiskt lexikon*, vol. 11. Stockholm: Albert Bonnier, pp. 48–53.

Möller, H. (2014) *Lyx och mode i stormaktstidens Sverige. Jesper Swedberg och kampen mot perukerna.* Stockholm: Atlantis.

Nadelmann, K. H. (1957) "On the origin of the bankruptcy clause" *The American Journal of Legal History.* 1(3), pp. 215–228.

Napolitano, M. R, Marino, V., & Ojala, J. (2015) "In search of integrated framework of business longevity" *Business History.* 57(7), pp. 955–969.

Nilsson, L. (1989) *Den urbana transitionen. Tätorterna i svensk samhällsomvandling 1800–1980.* Stockholm: Stadshistoriska institutet.

Nilsson, L. (1992) *Historisk tätortsstatistik. Del 1. Folkmängden i administrativa tätorter 1800–1970.* Stockholm: Stads- och kommunhistoriska institutet.

Nilzén, G. (2012) *Carl Gustaf Tessin: uppgång och fall.* Stockholm: Carlsson.

Nordin, J. (2013) *Versailles. Slottet, parken, livet.* Stockholm: Norstedt.

Nordin, J. G. (1881) *Handbok i boktryckarekonsten.* Stockholm: P. A. Norstedt & Söners Förlag.

Nordmark, D. (2001) "Liberalernas segertåg (1830–1858)", in Nordmark, D., Johannesson, E., & Petersson, B. (eds.), *Den svenska pressens historia.* vol. 2. Stockholm: Ekerlids förlag, pp. 18–125.

Nyberg, K. (1992) *Köpes: ull, säljes: kläde. Yllemanufakturens företagsformer i 1780-talets Stockholm.* Diss. Uppsala: Uppsala universitet.

Nyberg, K. (1999a) "Dagspressens roll i marknadsinstitutionens framväxt", in Gustafsson, K.-E. & Rydén, P. (eds.), *Folk och press i rörelse.* Göteborg: NORDICOM-Sverige, pp. 105–119.

Nyberg, K. (1999b) *Kommersiell kompetens och industrialisering. Norrköpings ylleindustriella tillväxt på Stockholms bekostnad 1780–1846.* Uppsala: Acta Universitatis Upsaliensis.

Nyberg, K. (2001) "The 'Skeppsbro nobility' in Stockholm's old town 1650–1850 a research program on the role and significance of trade capitalism in Swedish economy and society", in *Uppsala papers in economic history, Research Reports,* nr. 49. Uppsala: Uppsala universitet, pp. 1–21.

Nyberg, K. (2005) "Jag existerar endast genom att äga kredit". Tillit, kreditvärdighet och finansiella nätverk i 1700-talets och det tidiga 1800-talets Stockholm", in Berglund, M. (ed.), *Sakta vi gå genom stan:stadshistoriska studier tillägnade Lars Nilsson den 31/5 2005.* Stockholm: Stockholmia förlag, pp. 184–211.

Nyberg, K. (ed.) (2006) *Kopparkungen. Handelshuset Björkman i Stockholm 1782–1824.* Stockholm: Stockholmia förlag.

Nyberg, K. (2010) "The early modern financial system and the informal credit market", in Ögren, A. (ed.), *The Swedish Financial Revolution.* New York: Palgrave Macmillan, pp. 14–40.

Nyberg, K. (2015a) "Manufakturstatistikens källvärde", in Stavenow-Hidemark, E. & Nyberg, K. *Från kläde till silkesflor. Textilprover från 1700-talets svenska fabriker.* Stockholm: Kulturhistoriska Bokförlaget, pp. 321–324.

Nyberg, K. (2015b) "Anti-lyx. En samtida bildberättelse om Fersenska mordet den 20 juni 1810", in Wachenfeldt, P. von & Nyberg, K. (eds.), *Det svenska begäret. Sekler av lyxkonsumtion.* Stockholm: Carlssons bokförlag, pp. 130–164.

Nyberg, K. (2016) "Modets historiska ursprung. Den statliga manufakturens betydelse för modets uppkomst och spridning i Norden under tidigmodern tid", in Larsson, M., Palmsköld, A., Hörnfeldt, H., & Jönsson, L.-E. (eds.), *I utkanter och marginaler. 31 texter om kulturhistoria.* Stockholm: Nordiska museets förlag, pp. 338–350.

Nyberg, K. (ed.) (2017) *Ekonomisk kulturhistoria. Bildkonst, konsthantverk och scenkonst 1720–1850.* Stockholm: Kulturhistoriska bokförlaget.

Nyberg, K. & Ciszuk, M. (2015) "Textilmanufakturernas tillverkning 1744–1810", in Stavenow-Hidemark, E & Nyberg, K. *Från kläde till silkesflor. Textilprover från 1700-talets svenska fabriker.* Stockholm: Kulturhistoriska bokförlaget, pp. 325–348.

Nyberg, K. & Jakobsson, H. (2012) *Borgerskap och grosshandelssocietet i Stockholm. 1736–1850,* vol. 1. Stockholm: Informationsförlaget.

Nyberg, K. & Jakobsson, H. (2013) "Financial networks, migration and the transformation of the merchant elite in 18th–century Stockholm", in Safley, T. M. (ed.) *The history of bankruptcy. Economic, social and cultural implications in early modern Europe.* New York: Routledge, pp. 72–93.

Nyberg, K. & Jakobsson, H. (2016) "Negotiations, credit and trust in northern Europe. Institutional efficiency in the handling of bankruptcies in late eighteenth-century Stockholm", in Cordes, A. & Schulte Beerbühl, M. (eds.), *Dealing with economic failure. Between norm and practice (15th to 21st century).* Frankfurt am Main: Peter Lang, pp. 97–114.

Nyström, B. (2008) "Möbler över tiden – en inledning", in Nyström, B. & Eklund Nyström, S. (eds.), *Svenska möbler under femhundra år.* Stockholm: Natur & kultur.

Nyström, B., Biörnstad, A., & Bursell, B. (eds.) (1989) *Hantverk i Sverige.* Stockholm: LT:s förlag.

Nyström, P. (1955) *Stadsindustriens arbetare före 1800-talet: bidrag till kännedom om den svenska manufakturindustrien och dess sociala förhållanden.* Diss. Lund: Lunds universitet.

Odlinder, H. S. (2008) "Karl Johanstid", in Nyström, B. & Eklund Nyström, S. (eds.), *Svenska möbler under femhundra år.* Stockholm: Natur & kultur.

Ogilvie, S. (1997) *State corporatism and proto-industry. The Württemberg Black Forest, 1580–1797.* Cambridge: Cambridge University Press.

Ogilvie, S. (2003) *A bitter living. Women, markets, and social capital in early modern Germany.* Oxford: Oxford University Press.

Ogilvie, S. (2005) "The use and abuse of trust": social capital and its development by early modern guilds" *Jahrbuch für Wirtschaftsgeschichte.* 1, pp. 15–52.

Ogilvie, S. (2010) "Consumption, social capital, and the "industrious revolution" in early modern Germany" *The Journal of Economic History.* 70(2), pp. 287–325.

Ogilvie, S. (2019) *The European guilds. An economic analysis.* Princeton: Princeton University Press.

Ogilvie, S. C., Küpker, M., & Maegraith, J. (2012) "Household debt in early modern Germany. Evidence from personal inventories" *Journal of Economic History.* 72(1), pp. 134–167.

Ojala, J. (1997) "Approaching Europe: the merchant networks between Finland and Europe during the eighteenth and nineteenth centuries" *European Review of Economic History.* 1(3), pp. 323–352.

Olivecrona, S. R. (1862) *Bidrag till den svenska Concurslagstiftningens historia.* Uppsala: C. A. Leffler.

Olsson, M. (1932) "Jean Erik Rehns förbindelser med Manufakturkontoret under studieresan 1755–56", in Berg, G., Billow, A. & Selling, G. (eds.), *Gustavianskt.*

Studier kring den gustavianska tidens kulturhistoria tillägnade Sigurd Wallin på hans femtioårsdag. Stockholm: Nordisk rotogravyr, pp. 62–91.

Olsson, M. (1945) "Två dagböcker: från Jean Erik Rehns utrikesresa (1755–1756) och från John Tobias Sergels första italienska resa (1767–1779)" *RIG – Kulturhistorisk tidskrift.* 28, pp. 73–91.

Olsson, M. & Böttiger, J. (eds.) (1940–1941) *Stockholms slotts historia,* vols. 1–3 Stockholm: Norstedt.

Palat, R. A. (2015) *The making of an Indian ocean world-economy, 1250–1650. Princes, paddy fields, and bazaars.* New York: Palgrave Macmillan.

Pearson, R., Campbell, G., & Lemire, B. (eds.) (2001) *Women and credit. Researching the past, refiguring the future,* Oxford: Berg.

Perlinge, A. (2005) *Sockenbankirerna. Kreditrelationer och tidig bankverksamhet. Vånga socken i Skåne 1840–1900,* Diss. Stockholm: Stockholms universitet.

Perlinge, A. (2020) "Private wealth accumulation in eighteenth century Scania: intergenerational credit businesses and rural debt logic in Oppmanna" *Scandia.* 86(1), pp. 35–54.

Persson, C. (1993) *Stockholms klädesmanufakturer 1816–1846.* Diss. Stockholm: Stockholms universitet.

Poni, C. (1997) "Fashion as flexible production: the strategies of the Lyons silk merchants in the eighteenth century", in Sabel, C. F. & Zeitlin, J. (eds.), *World of possibilities. Mass production and flexibility in western industrialization.* Cambridge: Cambridge University Press, pp. 37–74.

Poukens, J. (2012) "'Tout-à-la-fois cultivateurs et commerçans': smallholder and the Industrious Revolution in eighteenth-century Brabant" *The Agricultural History Review.* 60(2), pp. 153–172.

Probst, J. (2019) *Growth, factor shares, and factor prices.* Lund: Lunds universitet.

Proschwitz, G. von (ed.) (1983) *Tableaux de Paris et de la cour de France 1739–1742. Lettres inédites de Carl Gustaf, comte de Tessin.* Göteborg: Acta Universitatis Gothoburgensis.

Proschwitz, G. von (ed.) (1995) *Carl Gustaf Tessin. Kulturpersonen och privatmannen, 1695–1770.* Stockholm: Atlantis.

Proschwitz, G. von (2002) *Carl Gustaf Tessin i Paris. Konst och politik: brevväxling med Carl Hårleman.* Stockholm: Norstedt.

Raj, K. (2007) *Relocating modern science. Circulation and the construction of knowledge in South Asia and Europe, 1650–1900.* Basingstoke: Palgrave Macmillan.

Raj, K. (2011) "The historical anatomy of a contact zone: Calcutta in the eighteenth century" *Indian Economic and Social History Review.* 48(1), pp. 55–82.

Rasmussen, P. (2010) *Skräddaren, sömmerskan och modet. Arbetsmetoder och arbetsdelning i tillverkningen av kvinnlig dräkt 1770–1830.* Diss. Uppsala: Uppsala universitet.

Rasmussen, P. (2019a) "Four *Robes de Cour* worn at the Swedish court: tradition and change in tailoring and significance", in Kammel, Frank Matthias & Pietsch, Johannes (eds.), *Structuring fashion. Foundation garments through history.* München: Hirmer, pp. 85–100.

Rasmussen, P. (2019b) "Frihetstidens hovdräkt: en närläsning av fyra "Robes de cour" i Livrustkammaren", in Karlsson, Klas-Göran, Severinsson, Emma & Zander, Ulf (eds.), *I historiens virvlar. Biografiska betraktelser tillägnade Eva Helen Ulvros.* Lund: Historiska Media, pp. 82–111.

Rebolledo-Dhuin, V. (2016) "Below and beyond bankruptcy—credit in the Parisian book trade in the nineteenth century", in Cordes, A. & Schulte Beerbühl, M. (eds.), *Dealing with economic failure. Between norm and practice (15th to 21st century)*. Frankfurt am Main: Peter Lang, pp. 139–174.

Reynard, P. C. (2001) "The language of failure. Bankruptcy in eighteenth-century France" *The Journal of European Economic History.* 30(2), pp. 355–390.

Ribeiro, A. (2004[1986/]) *Dress and morality.* Oxford: Berg.

Ribeiro, A. (2017) *Clothing art. The visual culture of fashion, 1600–1914.* New Haven: Yale University Press.

Richardson, C. (ed.) (2004) *Clothing culture, 1350–1650.* Burlington: Ashgate.

Riello, G. (2013) *Cotton. The fabric that made the modern world.* Cambridge: Cambridge University Press.

Riello, G. & McNeil, P. (eds.) (2010) *The fashion history reader. Global perspectives.* London: Routledge.

Riello, G. & Parthasarathi, P. (eds.) (2009) *The spinning world. A global history of cotton textiles, 1200–1850.* Oxford: Oxford University Press.

Riello, G. & Rublack, U. (eds.) (2019) *The right to dress. Sumptuary laws in a global perspective, 1200–1800.* New York: Cambridge University Press.

Riello, G. & Tirthankar, R. (eds.) (2009) *How India clothed the world. The world of south Asian textiles, 1500–1850.* Leiden: Brill.

Rimm, A.-M. (2009) *Elsa Fougt, Kungl. Boktryckare. Aktör i det litterära systemet ca 1780–1810.* Diss. Uppsala: Uppsala universitet.

Roberts, M. (1986) *The age of liberty. Sweden 1719–1772.* Cambridge: Cambridge University Press.

Roche. D. (1987[1981]) *The people of Paris. An essay in popular culture in the 18th century.* Berkeley & Los Angeles: University of California Press.

Roche, D. (1996) *The culture of clothing. Dress and fashion in the Ancien régime.* Cambridge & New York: Cambridge University Press.

Roche. D. (2000) *A history of everyday things. The birth of consumption in France, 1600–1800.* Cambridge: Cambridge University Press.

Rodger, R. G. (1985) "Business failure in Scotland 1839–1913" *Business History.* 27(1), pp. 75–99.

Roetzel, B. (1999) *Gentlemannen. Handbok i det klassiska herrmodet.* Köln: Könemann.

Roover, R. de (1948) *Money, banking and credit in mediaeval Bruges. Italian merchant-bankers Lombards and money-changers. A study in the origins of banking.* Massachusetts: Cambridge.

Ross, A. (ed.) (1974) *European bankruptcy laws.* Chicago: American Bar Association.

Rothstein, Natalie (1990) *Silk designs of the eighteenth century. In the collection of the Victoria and Albert Museum, London, with a complete catalogue.* London: Thames and Hudson.

Ruysscher, D. de (2013) "Bankruptcy, insolvency and debt collection among merchants in Antwerp (c. 1490–c.1540)", in Safley, T. M. (ed.) *The history of bankruptcy. Economic, social and cultural implications in early modern Europe.* New York: Routledge, pp. 185–199.

Ruysscher, D. de (2016) "The struggle for voluntary bankruptcy and debt adjustment in Antwerp (c. 1520–c. 1550)", in Cordes, Albrecht & Schulte Beerbühl. (eds.), *Dealing with economic failure. Between norm and practice (15th to 21st century)*. Frankfurt am Main: Peter Lang, pp. 77–95.

Rydin, K. (1888) *Om konkursförbrytelser enligt svensk rätt.* Uppsala: Almquist & Wiksell.

Sabel, Charles F. & Zeitlin, Jonathan (eds.) (1997) *World of possibilities. Mass production and flexibility in western industrialization.* Cambridge: Cambridge University Press.

Safley, T. M. (2000) "Bankruptcy: family and finance in early modern Augsburg" *The Journal of European Economic History.* 29(1), pp. 53–75.

Safley, T. M. (ed.) (2013) *The history of bankruptcy. Economic, social and cultural implications in early modern Europe.* New York: Routledge.

Saito, O. (2010) "An industrious revolution in an east Asian market economy? Tokugawa Japan and implications for the great divergence" *Australian Economic History Review.* 50(3), pp. 240–261.

Sallila, J. (2016) *Insolvency, commercial utility and principles of justice. The making of bankruptcy law in Sweden and Finland, ca. 1680–1868.* Diss. Helsinki: University of Helsinki.

Samuelsson, K. (1951) *De stora köpmanshusen i Stockholm 1730–1815. En studie i den svenska handelskapitalismens historia.* Diss. Stockholm: Stockholms högskola.

Sargentson, C. (1993) "The manufacture and marketing of luxury goods", in Brewer, J. & Porter, R. (eds.), *Consumption and the world of goods.* London: Routledge.

Sargentson, C. (1996) *Merchants and luxury markets. The marchands merciers of eighteenth-century Paris.* London: Victoria and Albert Museum.

Schulte Beerbühl, M. (2015) *The forgotten majority. German merchants in London, naturalization, and global trade, 1660–1815.* New York: Berghahn.

Schumpeter, J. (1989[1939]) *Business cycles.* New York: McGraw-Hill.

Schunka, A. (2005) "Glaubensflucht als Migrationsoption. Konfessionell motivierte Migrationen in der Frühen Neuzeit" *Geschichte in Wissenschaft und Unterricht.* 56(10), pp. 547–564.

Schäfer, D., Riello, G., & Molà, L. (eds.) (2018) *Threads of global desire. Silk in the pre-modern world.* Woodbridge: The Boydell Press.

Schön, L. (1979) *Från hantverk till fabriksindustri. Svensk textiltillverkning 1820–1870.* Diss. Lund: Arkiv.

Schön, L. (2006) *Tankar om cykler. Perspektiv på ekonomin, historia och framtiden.* Stockholm: SNS förlag.

Schön, L. (2017) *Sweden's road to modernity. An economic history.* Lund: Studentlitteratur.

Schön, L. & Krantz, O. (2012) "The Swedish economy in the early modern period. Constructing historical national accounts" *European Review of Economic History.* 16(4), pp. 529–549.

Seong Ho, J., Lewis, J. B., & Han-Rog, K. (2009) "Stability or decline? Demand or supply?" *The Journal of Economic History.* 69(4), pp. 1144–1151.

Sgard, J. (2006) "Do legal origins matter? The case of bankruptcy laws in Europe 1808-1914" *European Review of Economic History.* 10, pp. 389–419.

Sgard, J. (2013) "Bankruptcy, fresh start and debt renegotiation in England and France (17th–18th century)", in Safley, T. M. (ed.), *The history of bankruptcy. Economic, social and cultural implications in early modern Europe.* New York: Routledge, pp. 223–235.

Shaiman, S. L. (1960) "The history of imprisonment for debt and insolvency laws in Pennsylvania as they evolved from the common law" *The American Journal of Legal History.* 4(3), pp. 205–225.

Simonton, D. & Montenach, A. (eds.) (2013) *Female agency in the urban economy. Gender in European towns, 1640–1830.* New York: Routledge.

Simonton, D., Kaartinen, M., & Montenach, A. (eds.) (2015) *Luxury and gender in European towns, 1700–1914.* New York: Routledge.

Sjöberg, L. (1991) *Stolar, fåtöljer och soffor under empiren.* Stockholm: Nationalmuseum.

Skeel, D. A. (2001) *Debt's dominion. A history of bankruptcy law in America.* Princeton: Princeton University Press.

Skuncke, M.-C., & Tandefelt, H. (eds.) (2003) *Riksdag, kaffehus och predikstol. Frihetstidens politiska kultur 1766–1772.* Stockholm: Atlantis.

Smail, J. (1999) *Merchants, markets and manufacturers. The English wool textile industry in the eighteenth century.* Basingstoke: Macmillan.

Smail, J. (2005) "Credit, risk, and honor in eighteenth-century commerce" *Journal of British Studies.* 44(3), pp. 439–456.

Snodin, M. & Styles, J. (2004a) *Design & the decorative arts. Tudor and Stuart Britain 1500–1714.* London: V. & A. Publications.

Snodin, M. & Styles, J. (2004b) *Design & the decorative arts. Georgian Britain, 1714–1837.* London: V. & A. Publications.

Spence, C. (2016) *Women, credit, and debt in early modern Scotland.* Manchester: Manchester University Press.

Spufford, P. (2002) *Pengar och makt. Medeltidens handelsmän i Europa.* Stockholm: Bonniers.

Staf, N. (1940) *Marknadsreformen 1788. Till frågan om de enskilda marknadernas avskaffande.* Stockholm: Esselte.

Stavenow, Å. (1927) *Carl Hårleman. En studie i frihetstidens arkitekturhistoria.* Diss. Uppsala: Uppsala universitet.

Stavenow-Hidemark, E. & Nyberg, K. (2015) *Från kläde till silkesflor. Textilprover från 1700-talets svenska fabriker.* Stockholm: Kulturhistoriska bokförlaget.

Styles, J. (2007) *The dress of the people. Everyday fashion in eighteenth-century England.* New Haven: Yale University Press.

Sylvan, N. (1942) *Svensk realistisk roman 1795–1830.* Stockholm: Seelig.

Sylvén, T. (1996) *Mästarnas möbler. Stockholmsarbeten 1700–1850.* Stockholm: Norstedts.

Sylvén, T. (2003) *Stolens guldålder. Stolar och stolmakare i Sverige 1650–1850.* Stockholm: Prisma.

Sylwan, V. & Geijer, A. (1931) *Siden och brokader. Sidenväveriets och tygmönstrens utveckling: en översikt.* Stockholm: Natur och kultur.

Synnerberg, L. Nilsson (1815) *Svenskt waru-lexicon uti sammandrag ur de mest bekanta författares arbeten, rörande handeln.* 2 vols. Göteborg: Norberg.

Söderberg, J. (2007) *Vår världs ekonomiska historia.* vol. 1. Stockholm: SNS förlag.

Söderberg, J., Jonsson, U., & Persson, C. (1991) *A stagnating metropolis. The economy and demography of Stockholm 1750–1850.* Cambridge: Cambridge University Press.

Söderlund, E. (1943) *Stockholms hantverkarklass 1720–1772. Sociala och ekonomiska förhållanden.* Stockholm: Norstedts.

Söderlund, E. (1949) *Hantverkarna, del 2. Stormaktstiden, Frihetstiden och Gustavianska tiden.* Stockholm: Tiden.

Tétart-Vittu, F., Norberg, K., & Rosenbaum, S. L. (eds.) (2014) *Fashion prints in the age of Louis XIV. Interpreting the art of elegance.* Lubbock: Texas Tech University Press.

Thompson, E. L. (2004) *The reconstruction of southern debtors. Bankruptcy after the civil war.* Athens: University of Georgia Press.

Trentmann, Frank (ed.) (2012) *The Oxford handbook of the history of consumption.* Oxford: Oxford University Press.

Turunen, R. (2017) *Velka, vararikko ja tuomio. Konkurssi ja sen merkitykset 1800-luvun suomalaisissa kaupungeissa* [Debt, ruin and judgment. Bankruptcy and its meanings in nineteenth-century urban Finland]. Jyväskylä: University of Jyväskylä.

Turunen, R. (2018) "Kirjalliset velkasitoumukset 1800-luvun Suomessa" [Written debentures in nineteenth-century Finland], in *Ennen ja Nyt–Historian tietosanomat.* [Online publication: http://www.ennenjanyt.net/2019/02/kirjalliset-velkasitoumukset-1800-luvun-suomessa/, Cited 4 October 2019]

Tuula, M. (2001) *Rekonstruktion av företag inom insolvenslagstiftningens ramar. En jämförande studie av svensk och amerikansk insolvensrätt.* Stockholm: Stockholms universitet.

Udell, G. G. (ed.) (1968) *Bankruptcy laws of the United States,* Washington: U.S. Government Printing Office.

Ulväng, G. (2004) *Hus och gård i förändring. Uppländska herrgårdar, boställen och bondgårdar under 1700- och 1800-talens agrara revolution.* Hedemora: Gidlunds.

Ulväng, G. (2011) "Herrgårdsbyggandet i Mälardalen under 1700- och 1800-talet: när, var och av vem?" *Bebyggelsehistorisk tidskrift.* 60, pp. 38–57.

Ulväng, G., Murhem, S., & Lilja, K. (2013) *Den glömda konsumtionen. Auktionshandel i Sverige under 1700- och 1800-talen.* Hedemora: Gidlunds.

Utterström, G. (1954) "Some population problems in pre-industrial Sweden", *Scandinavian Economic History Review.* 2(2), pp. 101–165.

Van Zanden, J. L. (1999) "Wages and the standard of living in Europe, 1500–1800", *European Review of Economic History.* 3(2), pp. 175–197.

Vause, E. (2016) "The ties that bind? An analysis of the debt imprisonment records in Lyon 1835–40", in Cordes, A. & Schulte Beerbühl, M. (eds.), *Dealing with economic failure. Between norm and practice (15th to 21st century).* Frankfurt am Main: Peter Lang, pp. 175–191.

Vause, E. (2018) *In the red and in the black. Debt, dishonor, and the law in France between revolutions.* Charlottesville: University of Virginia Press.

Vilkuna, K. H. J. (2019) "Peruukki, puuteri ja varhaismoderni mies" [Wig, powder, and early modern man], in Turunen, A, & Niiranen, A. (eds.), *Säädyllistä ja säädytöntä. Pukeutumisen historiaa renessanssista 2000-luvulle.* Helsinki: Suomalaisen kirjallisuuden seura, pp. 88–114.

Vries, J. de (1994) "The industrial revolution and the industrious revolution" *The Journal of Economic History.* 54(2), pp. 249–270.

Vries, J. de (2008) *The industrious revolution. Consumer behavior and the household economy, 1650 to the present.* Cambridge: Cambridge University Press.

Wachenfeldt, P. von (2013) "The language of luxury in eighteenth-century France", Hancock II, J. H., Johnson-Woods, T., & Karaminas, V. (eds.), *Fashion in popular culture. Literature, media and contemporary studies.* Bristol: Intellect, pp. 209–223.

Wachenfeldt, P. von (2015) "I lyxens tjänst. Borgerliga värderingar och ideal i den moderna tiden", in Wachenfeldt, P. von & Nyberg, K. (eds.), *Det svenska begäret. Sekler av lyxkonsumtion.* Stockholm: Carlssons Bokförlag, pp. 164–188.

Wachenfeldt, P. von (2018) "The myth of luxury in a fashion world", in *Fashion, Style and Popular Culture.* 5(3), pp. 313–328.

Wachenfeldt, P. von & Nyberg, K. (eds.) (2015) *Det svenska begäret. Sekler av lyxkonsumtion.* Stockholm: Carlssons.

Wakefield, A. (2013) "The insolvent Zuchthaus as Cameralist Dystopia", in Safley, T, M. (ed.), *The history of bankruptcy. Economic, social and cultural implications in early modern Europe.* New York: Routledge, pp. 157–172.

Wallin, S. (1920) "Siden-droguet. En notis till 1700-talets svenska sidenfabrikation" *RIG – Kulturhistorisk tidskrift.* 3, pp. 53–62.

Walsh, C. (2014) "Stalls, bulks, shops and long-term change in seventeenth- and eighteenth-century England", in Furnée, J. H. & Lesger, C. (eds.), *The landscape of consumption. Shopping streets and cultures in Western Europe, 1600–1900.* Basingstoke: Palgrave Macmillan, pp. 37–56.

Weedon, A. (2013) "The economics of print", in Suarez, M. F. & Wuodhuysen, H. R. (eds.), *The book. A global history.* Oxford: Oxford University Press, pp. 154–168.

Weissbach, L. S. (1982) "Artisanal responses to artistic decline. The cabinetmakers of Paris in the era of industrialization" *Journal of Social History.* 16(2), pp. 67–81.

Welamson, L. (1961) *Konkursrätt.* Stockholm: Norstedt.

Welbourne, E. (1932) "Bankruptcy before the era of Victorian reform" *Cambridge Historical Journal.* 4(1), pp. 51–62.

Welch, E. S. (ed.) (2017) *Fashioning the early modern. Dress, textiles, and innovation in Europe, 1500–1800.* Oxford: Oxford University Press.

Wengström, G. (1925) "Fredrik Carl Boye af Gennäs", in Boëthius, B. (ed.), *Svenskt biografiskt lexikon,* vol. 5. Stockholm: Albert Bonniers Förlag, pp. 260–264.

Westerlund, L. (1988) *Provincialschäfrarna i Sverige åren 1739–66.* Åbo: Åbo akademi.

Wilmowsky, P. von. (2016) "Insolvency law: its roles and principles", in Cordes, A. & Schulte Beerbühl, M. (eds.), *Dealing with economic failure. Between norm and practice (15th to 21st century).* Frankfurt am Main: Peter Lang, pp. 243–260.

Wilson, R. G. (1973) "The supremacy of the Yorkshire cloth industry in the eighteenth century", in Harte, N. B. & Ponting, K. G. (eds.), *Textile history and economic history. Essays in honour of miss Julia de Lacy Mann.* Manchester: Manchester University Press, pp. 225–246.

Wolff, C. (2005) *Vänskap och makt. Den svenska politiska eliten och upplysnings-tidens Frankrike.* Helsingfors: Svenska litteratursällskapet i Finland.

Wottle, M. (2000) *Det lilla ägandet. Korporativ formering och sociala relationer inom Stockholms minuthandel 1720–1810.* Diss. Stockholm: Stockholms universitet.

Young, K. A. (1995) *Kin, commerce, community. Merchants in the port of Quebec 1717–1745.* New York: Peter Lang.

Zhan, S. (2019) *The land question in China. Agrarian capitalism, industrious revolution, and East Asian development.* London: Routledge.

Åbjörnsson, R. (2016) "Inledning", *Insolvensrättslig tidskrift.* 1, pp. 5–6.

Ågren, M. (1992) *Jord och gäld. Social skiktning och rättslig konflikt i södra Dalarna ca 1650–1850.* Diss. Uppsala: Acta Universitatis Upsaliensis.

Ågren, M. (1999) "Fadern, systern och brodern. Makt- och rättsförskjutningar genom 1800-talets egendomsreformer" *Historisk Tidskrift.* 4, pp. 683–708.

Ågren, M. (2009) *Domestic secrets. Women and property in Sweden, 1600–1857.* Chapel Hill: University of North Carolina Press.

Ågren, K. (2007) *Köpmannen i Stockholm. Grosshandlares ekonomiska och sociala strategier under 1700-talet.* Diss. Uppsala: Uppsala universitet.

Åmark, K. (1915) *Spannmålshandel och spannmålspolitik i Sverige 1719–1830.* Diss. Stockholm: Stockholms universitet.

Ögren, A. (ed.) (2010) *The Swedish Financial Revolution.* New York: Palgrave Macmillan.

Index

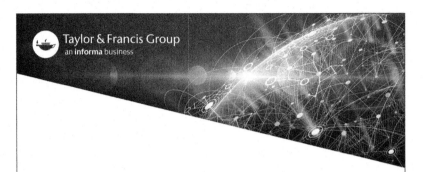

Printed in the United States
By Bookmasters